MATLAB® Primer

Eighth Edition

D1164452

MATLAB® Primer
Eighth Edition

Timothy A. Davis
University of Florida
USA

CRC Press
Taylor & Francis Group
Boca Raton   London   New York

CRC Press is an imprint of the
Taylor & Francis Group, an **informa** business

A CHAPMAN & HALL BOOK

CRC Press
Taylor & Francis Group
6000 Broken Sound Parkway NW, Suite 300
Boca Raton, FL 33487-2742

© 2011 by Taylor and Francis Group, LLC
CRC Press is an imprint of Taylor & Francis Group, an Informa business

No claim to original U.S. Government works

Printed in the United States of America on acid-free paper
10 9 8 7 6 5 4 3 2 1

International Standard Book Number: 978-1-4398-2862-5 (Paperback)

Library of Congress Cataloging-in-Publication Data

Davis, Timothy A.
    MATLAB primer / Timothy A. Davis. -- 8th ed.
        p. cm.
    Includes index.
    ISBN 978-1-4398-2862-5 (pbk. : alk. paper)
    1. MATLAB. 2. Numerical analysis--Data processing. I. Title.

QA297.D38 2011
518.0285--dc22   2010023858

Visit the Taylor & Francis Web site at
http://www.taylorandfrancis.com

and the CRC Press Web site at
http://www.crcpress.com

Contents

Preface

This eighth edition of the *MATLAB® Primer* highlights the new features of MATLAB 7.10 (R2010a), and expands on many existing features. New and expanded topics include:

- A new chapter on object-oriented programming.
- The MATLAB File Exchange window, which provides direct access to over 10,000 submissions by MATLAB users (as of March 2010).
- Major changes to the MATLAB Editor, such as code folding and the integration of the Code Analyzer (M-Lint) into the Editor.
- More powerful Help tools, such as quick help popups for functions via the Function Browser.
- The new bsxfun function.
- The Help chapter in the seventh edition gave a one-line description of every function, keyword, and operator. The number of functions and keywords has grown, and it has become impractical to keep up. This edition presents the *MATLAB Top 500*, and gives a longer synopsis of each of them. The list was determined via a MATLAB script that counted the occurrences of all functions and keywords in the entire File Exchange, with a few editorial modifications.
- Motivated by the MATLAB Top 500, several useful features not covered in the seventh edition have been added (such as sets, logical indexing, isequal, repmat, reshape, varargin, and varargout).

Tim Davis
Professor, Department of Computer and Information Science and Engineering, University of Florida,
www.cise.ufl.edu/~davis

Introduction

How to use this book: The purpose of this *MATLAB®Primer* is to help you begin to use MATLAB. Additional help is available inside MATLAB itself, and online at www.mathworks.com. The primer is best used hands on. You are encouraged to work at the computer as you read the primer and freely experiment with the examples. This primer, along with the MATLAB help facility, usually suffices for students in a class requiring the use of MATLAB.

Start with the examples at the beginning of each chapter. In this way, you will create all of the matrices and M-files used in the examples. Some examples depend on code you write in previous chapters and sections.

Larger examples (M-files and MEX-files) are posted on the web page for this book, at www.crcpress.com and www.cise.ufl.edu/~davis/MATLABPrimer8E.

Pull-down menu selections are described using the following style. Selecting the Desktop menu, and then the Desktop Layout submenu, and then the Default menu item is written as Desktop ▶Desktop Layout ▶Default.

MATLAB code and expressions are written in a fixed width font, like+this.

You should liberally use the online help facility for more detailed information. Selecting Help ▶Product Help brings up the Help window. You can also type help or doc in the Command Window. See Section 2.5 for more information on how to use the online help.

In the Help window, navigate to MATLAB ▶Functions. This gives you a categorical list of all functions, keywords, operators, and special characters in MATLAB. The outline of this list is repeated in Chapters A through K of the

Appendix of this book. Chapter L of the Appendix is an outline of the Symbolic Math Toolbox ►Functions categorical list. The Appendix describes the *Top 500* functions in MATLAB and The Symbolic Toolbox, which is a list of the most frequently used functions. Sometimes less is more, since you do not have to ponder over whether or not you need an obscure function when what you are looking for is a well-known and well-used function instead. In the interest of completeness, a few functions are described in the text of the book (Chapters 1-23), but which do not make it into the Top 500 list (linsolve is one example, in Section 5.5).

How to obtain MATLAB: Version 7.10 (Release R2010a) of MATLAB is available for Microsoft Windows®(XP, Server 2003 or 2008, Vista, and 7), Mac®(OS X 10.5.5 Leopard® and above 10.6.x Snow Leopard®), and most versions of Linux®. The Student Version of MATLAB includes MATLAB, Simulink®, the Symbolic Math Toolbox™, and six other Toolboxes. Everything discussed in this book can be done in the Student Version of MATLAB.

MATLAB®, Simulink®, and Handle Graphics®, are registered trademarks of The MathWorks, Inc. Symbolic Math Toolbox™is a trademark of The MathWorks, Inc. Mac®, MacBook®, Leopard®, and Snow Leopard® are registered trademarks of Apple, Inc. Linux® is a registered trademark of Linus Torvalds. UNIX® is a registered trademark of The Open Group. Ubuntu® is a registered trademark of Canonical, Inc.

For product information, please contact: The MathWorks, Inc., 3 Apple Hill Drive, Natick, MA, 01760-2098 USA. Tel: 508-647-7000. Fax: 508-647-7001. E-mail: info@mathworks.com. Web: www.mathworks.com.

1 Getting Started

MATLAB offers engineers, scientists, and mathematicians an intuitive language for expressing problems and their solutions mathematically and graphically. It integrates computation, visualization, and programming in a flexible, open environment. Complex numeric and symbolic problems can be solved in a fraction of the time required with other languages such as C, Fortran, or Java.

The *MATLAB Primer* is a hands-on introduction to this powerful tool developed by The MathWorks, Inc. Double-click the MATLAB icon to get started.

You can also launch MATLAB with the system command `matlab`. If you are running MATLAB across a network, it can be faster to run MATLAB without its desktop user-interface, using the `matlab -nodesktop` command. Not all MATLAB features are available if you use this option.

When you are finished, the `quit` or `exit` commands terminate MATLAB. You might be prompted to save any files you are editing. Before exiting, use the `save` command to save any variables in your workspace that you want to keep.

2 The MATLAB Desktop

MATLAB has an extensive graphical user interface. When MATLAB starts, the main MATLAB window appears,

containing several windows and menu bars. Not all
windows appear in the default configuration. The Desktop
menu controls the layout and appearance of the windows
and gives you a list of the windows you can use. This list is
shown below, alongside the sections and page numbers in
this book where they are discussed. The first four appear by
default, the first time you use MATLAB. If you reconfigure
your Desktop windows, MATLAB remembers what you
have modified and displays the same configuration the next
time you start MATLAB.

Command Window	Section 2.1	p. 3
Command History	Section 2.2	p. 7
Current Folder	Section 2.3	p. 7
Workspace	Section 2.4	p. 9
Help	Section 2.5	p. 10
Profiler	Section 11.7	p. 85
File Exchange	Section 2.6	p. 12
Editor	Section 6.1,	pp. 34,
	Section 6.2,	36,
	Section 11.2,	79,
	Chapter 23	163
Figures	Section 15.2	p. 107
Web Browser	Section 23	p. 163
Variable Editor	Section 2.7	p. 12
File and Folder Comparisons	Section 11.8	p. 85

The Start button in the bottom left corner of the MATLAB
Desktop brings up demos, tools, and other windows. Try
Start ►MATLAB ►Demos and run one of the demos from
the MATLAB Demo window.

All MATLAB windows are docked in the default desktop,
which means that they are tiled on the main MATLAB
window. You can undock a window by selecting the menu
item Desktop ► ⌐ Undock or by clicking its undock button:

Dock it with Desktop ► ↘ Dock... or the dock button:

↘

Close a window by clicking its close button:

✕

Reshape the window tiling by clicking on and dragging the window edges.

The menu bar at the top of the MATLAB window contains a set of buttons and pull-down menus for working with M-files, windows, preferences and other settings, web resources for MATLAB, and online MATLAB help. If a window is docked and selected, its menu bar appears at the top of the main MATLAB window.

2.1 Command Window

MATLAB expressions and statements are evaluated as you type them in the Command Window, and results of the computation are displayed there too. Expressions and statements are also used in M-files (more on this in Chapter 6). They are usually of the form:

```
variable = expression
```

or simply:

```
expression
```

Expressions are usually composed from operators, functions, and variable names. Evaluation of the expression produces a matrix (or other data type), which is then displayed on the screen or assigned to a variable for future use. If the variable name and = sign are omitted, a variable ans (for answer) is automatically created to which the result is assigned.

A statement is normally terminated at the end of the line. However, a statement can be continued to the next line with three periods (. . .) at the end of the line. Several statements can be placed on a single line separated by commas or semicolons. If the last character of a statement is a semicolon, display of the result is suppressed, but the assignment is still carried out. This is essential in suppressing unwanted display of intermediate results.

In the default configuration, the Workspace window in the top right of the MATLAB Desktop gives you a list of the variables you create. Type this command in the Command Window:

```
A = [1 2 3 ; 4 5 6 ; -1 7 9]
```

or this one:

```
A = [
1 2 3
4 5 6
-1 7 9]
```

Either one creates a 3-by-3 matrix and assigns it to a variable A. Try it. You will see the array A in your Workspace window (Section 2.4 gives more details on this window). MATLAB is case-sensitive in the names of commands, functions, and variables, so A and a are two different variables. A comma or blank separates the elements within a row of a matrix (sometimes a comma is necessary to split the expressions, because a blank can be ambiguous). A semicolon ends a row. When listing a number in exponential form (e.g., 2.34e-9), blank spaces must be avoided in the middle (before the e, for example). Matrices can also be constructed from other matrices. If A is the 3-by-3 matrix shown above, then:

4

```
C = [A, A' ; [12 13 14], zeros(1,3)]
```

creates a 4-by-6 matrix. Try it to see what C is. The quote mark in A' means the transpose of A. Be sure to use the correct single quote mark (just to the left of the enter or return key on most keyboards). Since a blank separates elements in a row, parentheses are sometimes needed around expressions if they would otherwise be ambiguous. Parentheses are also used for passing parameters to functions, such as the zeros function in this example. See Section 5.1 for more on the zeros function.

When you typed the last two commands, the matrices A and C were created and displayed in the Workspace window.

You can save the Command Window dialog with the diary command:

```
diary filename
```

This causes what appears subsequently in the Command Window to be written to the named file (if the filename is omitted, it is written to a default file named diary) until you type the command diary off; the command diary on causes writing to the file to resume. When finished, you can edit the file as desired and print it out. For hard copy of graphics, see Section 15.10.

The command line in MATLAB can be easily edited in the Command Window. The cursor can be positioned with the left and right arrows and the Backspace (or Delete) key used to delete the character to the left of the cursor.

A convenient feature is use of the up and down arrows to scroll through the stack of previous commands. You can recall a previous command line, edit it, and execute the revised line. Try this by first modifying the matrix A by adding one to each of its elements:

```
A = A + 1
```

You can change C to reflect this change in A by retyping the lengthy command C = ... above, but it is easier to hit the up arrow key until you see the command you want, and then hit enter.

Tab completion is another helpful shortcut. It works in both the Command Window and the Editor (see Section 6.1). Start typing a command name, a variable name, or a file name. Before you type it all in, hit the tab key. Try typing z then a tab. A list of all the functions and variables that start with z will pop up. Select one from the list, or keep typing to narrow down the selection. Type e and then tab to narrow down the selection to zeros.

Tab completion can be disabled in the Keyboard section of the File ▶Preferences menu. You can also use that menu to change your keyboard shortcuts.

You can clear the Command Window with the clc command or with Edit ▶Clear Command Window.

Beginning MATLAB users often wonder why MATLAB seems to compute its results in only 5 digits. Try this.

```
pi
```

No, MATLAB does not know a mere 5 digits of π. It keeps track of more digits than this, but only displays 5 digits by default. MATLAB typically does its computations in IEEE double precision floating point arithmetic, which is about 16 decimal digits. To see more digits, or to display numbers in different formats, try these commands:

format short	fixed point, 5 digits
format long	fixed point, 15 digits
format short g	fixed or scientific notation, 5 digits
format long g	fixed or scientific notation, 15 digits
format rat	approximate integer ratio

format short is the default. Once invoked, the chosen format remains in effect until changed. These commands only modify the display, not the precision of the number or its computation. To compute results in more digits you need to use variable precision arithmetic (Section 19.3).

The command `format compact` suppresses most blank lines, allowing more information to be placed on the screen. The command `format loose` returns to the non-compact format. These two commands are independent of the other format commands.

You can pause the output in the Command Window with the `more on` command. Type `more off` to turn this feature off.

2.2 Command History window

This window lists the commands typed in so far. You can re-execute one more commands from this window by double-clicking or dragging the command(s) into the Command Window. Try double-clicking on the command:

```
A = A + 1
```

shown in your Command History window. For more options, select and right-click on a line of the Command Window.

2.3 Current Folder window

The Current Folder window displays a list of the files in your current folder. The name of this folder also appears at the top of the main MATLAB window, in the MATLAB Toolbar. Your current folder is the first place MATLAB looks for your M-files, and for workspace (`.mat`) files containing data that you load and save. MATLAB also looks in all the folders in your MATLAB path (see Section 6.8). Folders that are not on your MATLAB path are shown in gray.

You can also load and save matrices as ASCII files and edit them with your favorite text editor. The file should consist of a rectangular array of just the numeric matrix entries. Use a text editor to create a file in your current folder called mymatrix.txt (or type edit mymatrix.txt) that contains these 2 lines:

```
22 67
12 33
```

Type the command load mymatrix.txt, and the file will be loaded from the current folder to the variable mymatrix. The file extension (.txt in this example) can be anything except .mat.

You can use the menus and buttons in the Current Folder window to peruse your files, or you can use commands typed in the Command Window. The command pwd returns the name of the current folder, and cd changes the current folder. Use cd .. to go to the parent folder. The command dir lists the contents of the current folder, whereas the command what lists only the MATLAB-specific files in the folder, grouped by file type. The MATLAB commands delete and type can be used to delete a file and display a file in the Command Window, respectively.

The Current Folder window can create and manage zip files. Right-click the mymatrix.txt file, and select Create Zip File. You can also create a zip file with multiple input files by selecting a set of files first. Double-clicking on the new mymatrix.zip file extracts its contents into a folder called mymatrix, containing the single file mymatrix.txt. Delete your original mymatrix.txt file. Clicking on the ⊞ symbol beside the mymatrix folder (or the ▶ symbol on the Mac) expands the contents of that folder. The name of the mymatrix.txt file it contains is grayed out, which tells you that MATLAB will not find that file if you type load `mymatrix.txt` (try it). Right-click the `mymatrix` folder and select Add to Path ►Selected Folders and try it again.

The Current Folder window includes a suite of useful code development tools for writing your own M-files. At this point in the book, you have yet to write your own M-files, so these tools are fully described later on (Chapter 11).

2.4 Workspace window

The Workspace window lists variables that you have either entered or computed in your MATLAB session.

There are many fundamental data types (or classes) in MATLAB, each one a multidimensional array. The classes you will use most are rectangular numerical arrays with possibly complex entries, and possibly sparse. An array of this type is called a matrix. A matrix with only one row or one column is called a vector (row vectors and column vectors behave differently; they are more than mere one-dimensional arrays). A 1-by-1 matrix is called a scalar.

Arrays can be introduced into MATLAB in several different ways. They can be entered as an explicit list of elements (as you did for matrix A), generated by statements and functions (as you did for matrix C), created in a file with your favorite text editor, or loaded from external data files or applications. You can also write your own functions (M-files and mexFunctions in C, Fortran, or Java) that create and operate on matrices. All the matrices and other variables that you create, except those internal to M-files, are shown in your Workspace window. Double-clicking on a variable in the Workspace window pulls up the Variable Editor (Section 2.7).

The command `whos` lists the variables currently in the workspace. Try typing `whos`; you should see a list of variables including A and C, with their type and size. A variable or function can be cleared from the workspace with the command clear *variablename* or by right-clicking the variable in the Workspace window and selecting Delete. The command clear alone clears all variables from the workspace.

When you log out or exit MATLAB, all variables are lost. However, invoking the command save before exiting writes all variables to a binary file named matlab.mat in the current folder. When you later reenter MATLAB, the command load restores the workspace to its former state. Commands save and load take file names and variable names as optional arguments. Type doc save and doc load, to bring up the documentation on these functions in the Help window described in the next section. Try typing the commands save, clear, and then load, and watch what happens in the Workspace window after each command.

2.5   Help window

This window is the most useful window for beginning MATLAB users, and you will continue to use it as you become an expert. Select Help ▶Product Help or type doc in the Command Window. The Help window has most of the features you would see in any web browser (clickable links, a back button, and a search tool, for example). The left panel shows where you are in the MATLAB online documentation. This book refers to Help sections in this window as Help:MATLAB ▶Getting Started ▶Introduction (for example), which means to select the MATLAB heading, then the Getting Started heading, and then the Introduction item under that heading. Clicking on the ⊞ symbol beside MATLAB in the left panel (or the ▶ symbol on the Mac) expands the MATLAB Contents.

Printable versions of the documentation are available from the Help:MATLAB page, under the heading *Printable (PDF) Documentation on the Web*. These are handy to download, read, and search when you are not running MATLAB, but you might hesitate to actually print them all out (they total nearly *12,000* pages in length). The *Getting Started Guide* is a gentle introduction to MATLAB and a mere 272 pages in length.

You can also use the `help` command, typed in the Command Window. For example, the command `help eig` tells about the eigenvalue function `eig`. See the list of functions in the Appendix for a brief summary of help for a function. `doc eig` shows you the full documentation of the `eig` function in the Help window.

The F1 key is a quick shortcut to getting help on a function. Inside the Command Window or Editor, after typing in a command, hit the F1 key. The Help window for that function will pop up.

For a quick index of all MATLAB functions, try the Function Browser. Select Help ►Function Browser (or type Shift-F1), and then drill down into one of the categories. For example, the `eig` function is found under MATLAB ►Mathematics ►Linear Algebra ►Eigenvalues and Singular Values ►`eig`. Selecting a function brings up a short description of the `eig` function, with a link for more help.

When you type a function name in the Command Window or in the Editor, followed by the left parenthesis, a small popup appears. Try typing `eig(`, but do not hit the Enter/Return key. The popup shows you the possible inputs to the function, and a link for more help.

You can also preview some of the features of MATLAB by first entering the command `demo` or by selecting Help ►Demos, and then selecting from the options offered. Most of the major features of MATLAB have their own demo. Some are videos, and some are interactive. Most demos are M-files that run step-by-step in the Command Window. The list of demos includes videos on the major new features for each release of MATLAB. These are very useful for keeping up-to-date on what MATLAB can do.

2.6   File Exchange window

The MathWorks, Inc., maintains a web site called MATLAB Central (www.mathworks.com/matlabcentral). It includes a Newsgroup, blogs, the Link Exchange, Webinars, programming contests, and the File Exchange.

The File Exchange is a place where any MATLAB user can post their MATLAB files for others to use. Quite often, if you want to solve a problem, someone else may have already solved it (avoid using this for homework solutions without your instructor's permission, of course). Users can rate the files, which helps you weed out the mediocre ones (a bad solution to a problem is worse than no solution at all).

With the File Exchange window, you can search for files from the File Exchange, download them, install them, and try them out. Select the Desktop ▶File Exchange menu option on the Desktop. If you do not have a MathWorks Account, you will be asked to create one. Try downloading the code that created the cover of this book by searching for "seashell" in the search box. Click on the green arrow to the right, download it, then type seashell in the Command Window. You can download all the codes in this book by searching for "MATLAB Primer."

2.7   Variable Editor window

Once an array or variable exists, it can be modified with the Variable Editor, which acts like a spreadsheet for matrices.

Go to the Workspace window and double-click on the matrix C. Click on an entry in C and change it, and try changing the size of C. Go back to the Command Window and type:

```
C
```

and you will see your new array C. You can also edit the matrix C by typing the command `openvar('C')`.

3 Matrices and Matrix Operations

You have now seen most of the windows in MATLAB and what they can do. Now take a look at how you can use MATLAB to work on matrices and other data types.

3.1 Referencing individual entries

Individual matrix and vector entries can be referenced with indices inside parentheses. For example, A(2,3) denotes the entry in the second row, third column of matrix A. Try:

```
A = [1 2 3 ; 4 5 6 ; -1 7 9]
A(2,3)
```

Next, create a column vector, x, with:

```
x = [3 2 1]'
```

or equivalently:

```
x = [3 ; 2 ; 1]
```

With this vector, x(3) denotes the third coordinate of vector x, with a value of 1. Higher dimensional arrays are similarly indexed. An array accepts positive integers as indices. You can also use logical indices, discussed in Section 5.8.

An array with two or more dimensions can be indexed as if it were a one-dimensional vector. If A is m-by-n, then A(i,j) is the same as A(i+(j-1)*m). This feature is most often used with the find function (see Sections 5.6 and 5.7).

3.2 Matrix operators

The following matrix operators are available in MATLAB:

+	addition or unary plus
-	subtraction or negation
*	multiplication
^	power
'	matrix transpose
.'	array transpose
\	left division (backslash or `mldivide`)
/	right division (slash or `mrdivide`)

These matrix operators apply, of course, to scalars (1-by-1 matrices) as well. If the sizes of the matrices are incompatible for the matrix operation, an error message will result, except in the case of scalar-matrix operations. With addition, subtraction, division, and multiplication of a matrix and a scalar, each entry of the matrix is operated on by the scalar, as in A=A+1. The scalar 1 is expanded in size to match the size of the matrix A. You can also expand non-scalars with the `bsxfun` function (see Section 5.9).

Not all scalar-matrix operations are valid. For example, `magic(3)/pi` is valid but `pi/magic(3)` is not. Also try the commands:

```
A^2
A*x
```

If x and y are both column vectors, then x'*y is their inner (or dot) product, and x*y' is their outer product. Try these commands:

```
y = [1 2 3]'
x'*y
x*y'
```

3.3 Matrix division (slash and backslash)

The matrix "division" operations deserve special comment. If A is an invertible square matrix and b is a compatible column vector or, respectively, a compatible row vector, then x=A\b is the solution of A*x=b, and x=b/A is the solution of x*A=b. These are informally called the backslash (\) and slash operators (/); they are also referred to as the mldivide and mrdivide functions. If A is square and non-singular, then A\b and b/A are mathematically the same as respectively, where inv(A) computes the inverse of A. The left and right division operators do not compute the inverse and are more accurate and efficient than inv(A)*b. In left division, if A is square, then it is factorized (if necessary), and these factors are used to solve A*x=b. If A is not square, the under- or over-determined system is solved in the least squares sense. Right division is defined in terms of left division by b/A=(A'\b')'. Try this:

```
A = [1 2 ; 3 4]
b = [4 10]'
x = A\b
```

The solution to A*x=b is the column vector x=[2;1]. Backslash is a very powerful general-purpose method for solving linear systems. Depending on the matrix, it selects forward or back substitution for triangular matrices (or permuted triangular matrices), Cholesky factorization for symmetric matrices, LU factorization for square matrices, or QR factorization for rectangular matrices. It has a special solver for Hessenberg matrices. It can also exploit sparsity, with either sparse versions of the above list, or special-case solvers when the sparse matrix is diagonal, tridiagonal, or banded. It selects the best method automatically (sometimes trying one method and then another if the first method fails).

This can be overkill if you already know what kind of matrix you have. It can be much faster to use the `linsolve` function described in Section 5.5.

3.4 Entry-wise operators

Matrix addition and subtraction already operate entry-wise, but the other matrix operations do not. These other operators (`*`, `^`, `\`, and `/`) can be made to operate entry-wise by preceding them by a period. For example, either:

```
[1 2 3 4] .* [1 2 3 4]
[1 2 3 4] .^ 2
```

yields `[1 4 9 16]`. Try it. This is particularly useful when using MATLAB graphics. Also compare `A^2` with `A.^2`.

3.5 Relational operators

The relational operators in MATLAB are:

`<`	less than
`>`	greater than
`<=`	less than or equal
`>=`	greater than or equal
`==`	equal
`~=`	not equal

They all operate entry-wise. Note that `=` is used in an assignment statement whereas `==` is a relational operator. Relational operators may be connected by logical operators:

`&`	and
`|`	or
`~`	not
`&&`	short-circuit and
`||`	short-circuit or

16

The result of a relational operator is of type `logical`, and is either `true` (one) or `false` (zero). Thus, ~0 is 1, ~3 is 0, and 4&5 is 1, for example. When applied to scalars, the result is a scalar. Try entering 3<5, 3>5, 3==5, and 3==3. When applied to matrices of the same size, the result is a logical matrix of ones and zeros giving the value of the expression between corresponding entries. You can also compare elements of a matrix with a scalar. Try:

```
A = [1 2 ; 3 4]
A >= 2
B = [1 3 ; 4 2]
A < B
```

The short-circuit operator && acts just like its non-short-circuited counterpart (&), except that it evaluates its left expression first, and does not evaluate the right expression if the first expression is false. This is useful for partially-defined functions. Suppose f(x) returns a logical value but generates an error if x is zero. The expression (x~=0) && f(x) returns false if x is zero, without calling f(x) at all. The short-circuit or (||) acts similarly. It does not evaluate the right expression if the left is true. Both && and || require their operands to be scalar and convertible to logical, while & and | can operate on arrays.

3.6 Complex numbers

MATLAB allows complex numbers in most of its operations and functions. Three convenient ways to enter complex matrices are:

```
clear i
B = [1 2 ; 3 4] + i*[5 6 ; 7 8]
B = [1+5i, 2+6i ; 3+7i, 4+8i]
B = complex([1 2 ; 3 4], [5 6 ; 7 8])
```

Either i or j may be used as the imaginary unit. You can use i and j as variables and overwrite their values, since they are also commonly used as loop indices (this is why the example above starts with `clear i`). You may generate a new imaginary unit with, say, `ii=sqrt(-1)`. The simplest thing to do is to always use the constants `1i` or `1j`, which cannot be reassigned and are always equal to the imaginary unit. Thus,

```
B = [1 2 ; 3 4] + 1i*[5 6 ; 7 8]
```

generates the same matrix B, even if i has been reassigned. See Section 10.2 for how to find out if i has been reassigned.

3.7 Strings

Enclosing text in single quotes forms strings with the `char` data type:

```
S = 'I love MATLAB'
```

To include a single quote inside a string, use two of them together, as in:

```
S = 'Green''s function'
```

A 2-D array of strings can represent multiple lines of text. For example,

```
S = [ 'I love MATLAB'
      'It''s powerful' ]
```

Then S(1,:) is the first line of text and S(2,:) is the second (colon notation is discussed in the next chapter).

Strings, numeric matrices, and all other data types can be displayed with the function `disp`. Try `disp(S)` and `disp(B)`.

Since all rows in an array must have the same number of entries, the strings must all be the same length, so you must pad shorter strings with spaces. Cell arrays avoid this problem (see Section 8.1).

Use strcmp and strcmpi to compare strings for equality. strcmp('A','a') is false, while strcmpi('A','a') is true because the latter ignores case.

To convert a number displayed a string into a number, use str2double, str2num, or sscanf. str2double('3.14') is the number 3.14. The str2num function can extract multiple numbers from a single string, but it it evaluates the string as if it were a MATLAB expression. The string might include calls to a function, which may cause unintended side effects. The sscanf function provides more more control over how numbers are parsed from strings. See also fscanf for reading numbers from text files.

4   Submatrices and Colon Notation

Vectors and submatrices are often used in MATLAB to express simple yet powerful matrix computations and data manipulations. Colon notation (which is used to both generate vectors and reference submatrices) and subscripting by integral vectors are keys to efficient manipulation of these objects. Creative use of these features minimizes the use of loops (which can slow down MATLAB) and makes code simple and readable. Make a special effort to become familiar with them.

4.1   Generating vectors

The expression 1:5 is the row vector [1 2 3 4 5]. The numbers need not be integers, and the increment need not be 1. For example, `0:0.2:1` gives `[0 0.2 0.4 0.6 0.8 1]` with an increment of `0.2` and `5:-1:1` gives `[5 4 3 2 1]` with an increment of `-1`. These vectors are commonly used in `for` loops, described in Section 7.1. Be careful how you mix the colon operator with other operators. Compare `1:5-3` with `(1:5)-3`.

In general, the expression `lo:hi` is the sequence `[lo, lo+1, lo+2, ..., hi]` except that the last term in the sequence is always less than or equal to `hi` if either one are not integers. Thus, `1:4.9` is `[1 2 3 4]` and `1:5.1` is `[1 2 3 4 5]`. The sequence is empty if `lo>hi`.

If an increment is provided, as in `lo:inc:hi`, the sequence is `[lo, lo+inc, lo+2*inc, ..., lo+m*inc]` where `m=fix((hi-lo)/inc)` and `fix` is a function that rounds a real number towards zero. The length of the sequence is `m+1`, and the sequence is empty if `m<0`. Thus, the sequence `5:-1:1` has `m=4` and is of length 5, but `5:1:1` has `m=-4` and is thus empty. The default increment is 1.

If you want specific control over how many terms are in the sequence, use `linspace` instead of the colon operator. The expression `linspace(lo,hi)` is identical to `lo:inc:hi`, except that `inc` is chosen so that the vector always has exactly 100 entries (even if `lo` and `hi` are equal). The last entry in the sequence is always `hi`. To generate a sequence with n terms instead of the default of 100, use `linspace(lo,hi,n)`. Compare `linspace(1,5.1,5)` with `1:5.1`.

4.2 Accessing submatrices

Colon notation can be used to access submatrices of a matrix. To try this out, first type the two commands:

```
A = rand(6,6)
B = rand(6,4)
```

which generate a random 6-by-6 matrix A and a random 6-by-4 matrix B.

A(1:4,3) is the column vector consisting of the first four entries of the third column of A.

A colon by itself denotes an entire row or column: A(:,3) is the third column of A, and A(1:4,:) is the first four rows of A.

Arbitrary integral vectors can be used as subscripts. A(:,[2 4]) is a matrix with two columns: columns 2 and 4 of A. This subscripting can be used on both sides of an assignment statement:

```
A(:,[2 4 5]) = B(:,1:3)
```

replaces columns 2,4,5 of A with the first three columns of B. Try it. The entire altered matrix A is displayed because the statement is not terminated with a semicolon. Columns 2 and 4 of A can be multiplied on the right by the matrix [1 2 ; 3 4]:

```
A(:,[2 4]) = A(:,[2 4]) * [1 2 ; 3 4]
```

Once again, the entire altered matrix is displayed. Submatrix operations are a convenient way to perform many useful computations. For example, a Givens rotation of rows 3 and 5 of the matrix A to zero out the A(3,1) entry can be written as:

```
a = A(5,1)
b = A(3,1)
G = [a b ; -b a] / norm([a b])
A([5 3], :) = G * A([5 3], :)
```

(assuming norm([a b]) is not zero). You can also assign a scalar to all entries of a submatrix. Try:

```
A(:, [2 4]) = 99
```

21

You can delete rows or columns of a matrix by assigning the empty matrix ([]) to them. Try:

```
A(:, [2 4]) = []
```

In an array index expression, end denotes the index of the last element. Try:

```
x = rand(1,5)
x = x(end:-1:1)
```

MATLAB is a powerful and expressive language. To appreciate the usefulness of these features, compare these MATLAB statements with the equivalent code in C, Fortran, or Java.

5 MATLAB Functions

MATLAB has a wide assortment of built-in functions. You have already seen some of them, such as zeros, rand, and sqrt. This chapter describes the more common matrix manipulation functions. For a more complete list, see the Appendix (p. 165), or Help:MATLAB ▶Functions.

5.1 Constructing matrices

Convenient matrix building functions include:

eye	identity matrix
zeros	matrix of zeros
ones	matrix of ones
diag	create or extract diagonals
triu	upper triangular part of a matrix
tril	lower triangular part of a matrix
rand	randomly generated matrix
hilb	Hilbert matrix
magic	magic square
toeplitz	Toeplitz matrix
gallery	a wide range of interesting matrices

The command `rand(n)` creates an n-by-n matrix with randomly generated entries distributed uniformly between 0 and 1 while `rand(m,n)` creates an m-by-n matrix (m and n are non-negative integers). Try:

```
A = rand(3)
```

Use `reset(RandStream.getDefaultStream)` to reset the random number generator, which is useful for reproducing the results from `rand`. Use `randn` for random numbers with normal distribution, and `randi` for random integers. All three functions (`rand`, `randn`, and `randi`) use the same underlying generator. `randi([0 9],3)` creates a random 3-by-3 matrix with integer entries in the range 0 to 9.

`zeros(m,n)` produces an m-by-n matrix of zeros, and `zeros(n)` produces an n-by-n one. If A is a matrix, then `zeros(size(A))` produces a matrix of zeros having the same size as A. The `ones` function is the same as `zeros`, except that it returns a matrix of all ones.

For a regular scalar, vector, or matrix, `size(A)` returns a vector of length 2 with the number of rows and columns of A For a scalar x, `size(x)` is [1 1]. With two outputs (or more) the dimensions of A are returned in the two (or more) scalars. A matrix A is *empty* if `min(size(A)` is zero; `isempty(A)` is true for an empty matrix A.

If x is a vector, `diag(x)` is the diagonal matrix with x down the diagonal; if A is a matrix, then `diag(A)` is a vector consisting of the diagonal of A. Try:

```
x = 1:3
diag(x)
diag(A)
diag(diag(A))
```

Matrices can be built from blocks. Try creating this 5-by-5 matrix.

```
B = [A zeros(3,2) ; pi*ones(2,3), eye(2)]
```

`eye(2)` is the 2-by-2 identity matrix. `magic(n)` creates an n-by-n matrix that is a magic square (rows, columns, and diagonals have common sum). Matrices can also be generated with a `for` loop (see Section 7.1). `triu` and `tril` extract upper and lower triangular parts of a matrix. Try:

```
triu(A)
triu(A) == A
```

The `gallery` function can generate a matrix from any one of over 60 different matrix classes. Many have interesting eigenvalue or singular value properties, provide interesting counter-examples, or are difficult matrices for various linear algebraic methods. The Parter matrix has many singular values close to π:

```
A = gallery('parter', 6)
svd(A)
```

Additional test matrices are available in other functions. For example, the Rosser matrix challenges many eigenvalue solvers:

```
A = rosser ;
eig(A)
eigs(A)
```

The `eig`, `eigs`, and `svd` functions are discussed in Section 5.4.

5.2 Scalar functions

Certain MATLAB functions operate essentially on scalars but operate entry-wise when applied to a vector or matrix. The most common ones are:

abs	atan2	exp	log10	rem	sqrt
acos	ceil	floor	log2	round	tan
asin	conj	imag	mod	sign	
atan	cos	log	real	sin	

The following statements generate a sine table:

```
x = (0:0.1:2)'
y = sin(x)
[x y]
```

Note that because `sin` operates entry-wise, it produces a vector y from the vector x.

5.3   Vector functions and data analysis

Other MATLAB functions operate essentially on a vector (row or column) but act on an m-by-n matrix (m > 2) in a column-by-column fashion to produce a row vector containing the results of their application to each column. Row-by-row action can be obtained by using the transpose (mean(A')', for example) or by specifying the dimension along which to operate (mean(A,2), for example). Most of these functions perform basic statistical computations (`std` computes the standard deviation and `prod` computes the product of the elements in the vector, for example). The primary functions are:

| max | sum | median | any | sort | var |
| min | prod | mean | all | std | mode |

The maximum entry in a matrix A is given by max(max(A)) rather than max(A). Try it. The `any` and `all` functions are discussed in Section 7.6.

5.4 Matrix functions

Much of the power of MATLAB comes from its matrix functions. Here is a partial list of the most important ones:

eig	eigenvalues and eigenvectors
eigs	like eig, for large sparse matrices
chol	Cholesky factorization
svd	singular value decomposition
svds	like svd, for large sparse matrices
lu	LU factorization
qr	QR factorization
poly	characteristic polynomial
det	determinant
size	size of an array
length	length of a vector
norm	1-norm, 2-norm, Frobenius-norm, ∞-norm
cond	condition number in the 2-norm
condest	condition number estimate
rank	rank
kron	Kronecker tensor product
find	find indices of nonzero entries
linsolve	solve a special linear system

MATLAB functions may have single or multiple output arguments. Square brackets are used to the left of the equal sign to list the outputs. For example,

    lambda = eig(A)

finds a column vector with the eigenvalues of A, whereas:

    [V, Lambda] = eig(A)

produces a matrix V whose columns are the eigenvectors of A and a diagonal matrix D with the eigenvalues of A on its diagonal, so that A*V is equal to V*Lambda. Try it.

5.5 The linsolve function

The matrix divide operators (\ or /) are usually enough for solving linear systems. They look at the matrix and try to pick the best method. The linsolve function acts like \, except that you can tell it about your matrix. Try:

```
A = [1 2 ; 3 4]
b = [4 10]'
A\b
linsolve(A,b)
```

In both cases, you get solution x=[2;1] to the linear system A*x=b.

If A is symmetric and positive definite, one explicit solution method is to perform a Cholesky factorization, followed by two solves with triangular matrices. Try:

```
C = [2 1 ; 1 2]
x = C\b
```

Here is an equivalent method:

```
R = chol(C)
y = R'\b
x = R\y
```

The matrix R is upper triangular, but MATLAB must explicitly determine this by itself. You can save MATLAB some work by using linsolve with an optional third argument, opts. Try this:

```
opts.UT = true
opts.TRANSA = true
y = linsolve(R,b,opts)
```

which gives the same answer as y=R'\b. The fields for the opts struct are UT (upper triangular), LT (lower triangular), UHESS (upper Hessenberg), SYM (symmetric), POSDEF

(positive definite), RECT (rectangular), and TRANSA
(whether to solve A*x=b or A'*x=b). All opts fields are
either true or false. Not all combinations are supported
(type doc linsolve for a list). linsolve does not work
on sparse matrices.

This example uses a struct called opts. Structs are covered
in detail in Section 8.2.

5.6   The find function

The find function is unlike the other matrix and vector
functions. find(x), where x is a vector, returns an array of
indices of nonzero entries in x. This is often used in
conjunction with relational operators. Suppose you want a
vector y that consists of all the values in x greater than 1.
Try:

```
x = 2*rand(1,5)
y = x(find(x > 1))
```

With three output arguments, you get more information:

```
A = rand(3)
[i,j,x] = find(A)
```

returns three vectors, with one entry in i, j, and x for each
nonzero in A (row index, column index, and numerical
value, respectively). With this matrix A, try:

```
[i,j,x] = find(A > .5)
[i j x]
```

and you will see a list of pairs of row and column indices
where A is greater than .5. However, x is a vector of values
from the matrix expression A > .5, not from the matrix A.
Getting the values of A that are larger than .5 without a
loop requires one-dimensional array indexing.

5.7 1-D indexing and the reshape function

One-dimensional or linear indexing accesses a multi-dimensional array with a single index.

```
A = rand(5)
k = find(A > .5)
A(k)
A(k) = A(k) + 99
```

Section 7.1 shows the loop-based version of this code.

If A is m-by-n, then A(i,j) is the same as the one-dimensional index A(i+(j-1)*m). Think of one-dimensional indexing starting at A(1,1), which is the same as A(1), and then counting down the first column. So A(m,1) is the same as A(m). Next, count down the second column, so that A(1,2) is the same as A(m+1). Try this example, where m=3.

```
A = rand(3)
A(1,2)
A(4)
```

A single colon converts a matrix into a vector, one column following another. Try this:

```
A(:)
```

One-dimensional indexing works differently depending on which side of the assignment (=) it is on. On the right of the equal sign, A(k) returns the values given by the indices in k. The result is the same size and shape as k. Try these examples:

```
A = magic(5)
A([1 2 ; 5 6])
A(1:7)
A([1:7]')
```

On the left side of the equal sign, A is modified but it retains its shape and size. The result of A(k)=... does not depend on the shape of k, just the number of entries it contains. If k is a matrix or row vector, it is converted into the single column k(:). Try this:

```
A([1 2 ; 5 6]) = 101:104
A(1:7) = 0
A([1:7]') = -1
```

The example above where k is 2-by-2 may seem out of order until you see it expanded like this:

```
A = magic(5)
k = [1 2 ; 5 6]
k = k(:)
A(k) = 101:104
```

The expression A(:) unravels A into a single column vector, starting with entries in the first column, then the second, and so on. The resulting column vector can then be reshaped into a matrix with different dimensions using reshape. C=reshape(A,m,n) unravels A(:) and then reshapes the result into an m-by-n matrix C. Try this example.

```
A = reshape(1:16, 4, 4)
C = reshape(A, 2, 8)
```

5.8 Logical indexing

Logical indexing is an alternative to one-dimensional indexing and the find function. You can index into an array using a list of indices (such as those from find), or you can index into it with a logical array directly. If S is a logical array, A(S) is the set of entries in A where S is true. Here is the same example as the first one in Section 5.7, but using logical indexing instead of find.

```
A = rand(5)
S = A > .5
A(S)
A(S) = A(S) + 99
```

The logical array S contains values that are true (1) or false (0). The expression x=A(S) returns a list x of values A(i,j) where S(i,j) is 1. S can be the same size as A, or smaller. Unlike one-dimensional indexing, the result x is always a vector and never a matrix. If S is a matrix or column vector, then x=A(S) is a column vector. If S is a row vector, then so is x=A(S). If S is smaller than A, S(:) gets padded with zeros to give it the same number of entries as A. S cannot be larger than A.

Merely creating a matrix with 1's and 0's does not make it a logical array. Try this example. S is an array of type double, not logical, so A(S) fails. You must convert it into a logical array with S~=0, or logical(S).

```
S = randi([0 1], 5)
A(S)        % fails
A(S==1)     % OK
```

5.9 The bsxfun and repmat functions

Scalars, vectors, and matrices work together in specific ways according to the well-defined rules of linear algebra. Entry-wise operations (such as adding two matrices A+B) can only be done if A and B have the same size. MATLAB can implicitly expand a scalar so that it takes on the same size as a matrix, as in A+1. This example subtracts a scalar (the mean of all entries in A) from the whole matrix:

```
A = rand(5)
C = A - mean(A(:))
```

31

Sometimes these rules are too restrictive. For example, subtracting the mean of each column from a matrix is a common operation in statistics. MATLAB cannot do that directly (A-mean(A) fails, even though the meaning of that expression is clear).

The repmat function is one approach for solving this problem. repmat(A,m,n) replicates and tiles the matrix A, m times along the rows and n times along the columns. repmat(A,1,2) is [A A] and repmat(A,2,1) is [A;A]. repmat(A,2,3) results in [A A A;A A A], To subtract the mean of each column from A, you can use repmat to replicate the row vector mean(A).

```
n = size(A,1)
C = A - repmat(mean(A), n, 1)
```

repmat is useful in its own right, but a better solution to this problem is bsxfun (short for Binary Singleton eXpansion Function).

C=bsxfun(f,A,B) applies the function f entry-wise to each pair of entries in A and B. See Sections 3.4 and 3.5 for most of the entry-wise functions in MATLAB. The dimensions of A and B must either match, or either can be equal to one (a singleton dimension). A singleton dimension in one matrix is expanded (implicitly replicating the matrix) to match a non-singleton dimension in the other matrix.

This example subtracts the mean of A from each column. All the columns of the new matrix C have a mean of zero (or nearly so, ignoring floating-point roundoff).

```
C = bsxfun(@minus, A, mean(A))
mean(C)
```

Here is how the above example works. x=mean(A) returns a row vector with x(k) equal to the mean of A(:,k). If A is m-by-n, then x is 1-by-n, so bsxfun expands the first

32

dimension by implicitly replicating the row vector x into a matrix of m rows. Subtracting this from A subtracts the mean of column k from A(:,k), for each column k.

To subtract the mean of row k from the kth row A(k,:), for each row k, try this:

```
C = bsxfun(@minus, A, mean(A,2))
mean(C,2)
```

The @ symbol creates a *function handle*, which is described in more detail in Section 10.1. Each MATLAB operator has a function name associated with it. You must add @ to the name to create a function handle, which can then be passed to bsxfun. Here is a list of the most common entry-wise functions to use with bsxfun (see doc bsxfun for a complete list):

@plus	+	@minus	-
@times	.*	@power	.^
@rdivide	./	@ldivide	.\
@lt	<	@gt	>
@le	<=	@ge	>=
@eq	==	@ne	~=
@max	maximum	@min	minimum

For example, to subtract the smallest entry of each column k from A(:,k), try this:

```
C = bsxfun(@minus, A, min(A))
min(C)
```

Another application of bsxfun is row equilibration, which scales each row of a matrix by dividing it by the largest entry in that row. It can be written as $C = D^{-1}A$ where D is a diagonal matrix with $d_{ii} = max_j|a_{ij}|$. This gives each row of C the same ∞-norm, which can improve the solution of a linear system. After row equilibration, the largest entry in each row of C has a magnitude of 1. Here are three ways of doing it. The first uses bsxfun, and the other two use backslash with a full and sparse scaling matrix D, respectively.

```
d = max(abs(A), [], 2)
C = bsxfun(@ldivide, d, A)
C = diag(d) \ A
n = size(A,1)
C = spdiags(d,0,n,n) \ A
```

For large matrices, the second method is much slower than the other two. The advantage of bsxfun is that it can handle many kinds of matrix operations that backslash cannot handle. Sparse matrices are discussed in Chapter 18.

6   M-Files

Most of the examples you have typed in so far are short. For solving more complex problems, you need to create a sequence of statements stored in a file so that you can easily edit and re-use them. These are called M-files because they must have the file type .m as the last part of their filename.

6.1   M-File Editor window

Much of your work with MATLAB will be in creating and refining M-files. M-files are usually created using with the M-file Editor or your favorite text editor. See also Help:MATLAB ▶ User Guide ▶ Desktop Tools and Development Environment ▶ Editing and Debugging M-Files. There are two types of M-files: script files and function files. In this exercise, you will incrementally develop and debug a script and then a function for making a matrix diagonally dominant. A matrix is diagonally dominant if the absolute value of each diagonal is larger than the sum of the absolute values of the off-diagonal entries in its row. That is, $|a_{ii}| > \sum_{j \neq i} |a_{ij}|$ for all i.

Create a new M-file, either with the `edit` command, by selecting the File ►New ►Script menu item, or by clicking the new-file button:

Type in these lines in the Editor.

```
f = sum(A, 2) ;
A = A + diag(f) ;
```

The semicolons are there because you normally do not want to see the results of every line of a script or function. Save the file as `ddom.m` by clicking:

You have just created your first MATLAB script file.

See the web page for this book for the M-files and MEX-files used in this book.

You might find it convenient to dock the Editor window (click ➘) because in the exercise below you will be going back and forth between the Editor window and the Command Window.

The Editor has many features that are introduced slowly in this book as you build up experience in MATLAB. More discussion on some of the advanced features of the Editor is found in Section 11.2 (*Advanced Editor features*) and Chapter 23 (*Cell Publishing*). The next section describes how to debug your code via breakpoints in the Editor.

6.2 Script files

A script file consists of a sequence of MATLAB statements that you could also type directly into the Command Window. Typing ddom in the Command Window causes the statements in the script file ddom.m to be executed. Variables in a script file refer to variables in the main workspace, so changing them changes your workspace variables. Type:

```
A = rand(3)
ddom
A
```

in the Command Window. It seems to work; the matrix A is now diagonally dominant. If you type this in the Command Window, though,

```
A = [1 -2 ; -1 1]
ddom
A
```

then the diagonal of A just got worse. What happened? Look at f in the Workspace window (or double-click on it to open f for editing); it is a column vector with the values [-1 ; 0]. Oops. f is supposed to be a sum of absolute values, so it cannot be negative. Change the first line of ddom.m to:

```
f = sum(abs(A), 2) ;
```

save the file, and run it again on the original matrix A=[1 -2;-1 1] (hit the up arrow key, or double-click the command in your Command History window). This time, instead of typing in the command, try running the script by clicking:

in the Editor window. This is a shortcut to typing ddom in the Command Window. The matrix A is now diagonally dominant. Run the script again, though, and you will see that A is modified even if it is already diagonally dominant. Fix this by modifying only those rows that violate diagonal dominance.

Set A to [1 -2;-1 1] again and modify ddom.m so that it looks like this:

```
d = diag(A) ;
a = abs(d) ;
f = sum(abs(A), 2) - a ;
i = find(f >= a) ;
A(i,i) = A(i,i) + diag(f(i)) ;
```

Save and run the script by clicking:

Run it again. This time the matrix does not change.

However, if you try it on the matrix A=[-1 2;1 -1], the result is wrong. To fix it, try another debugging method: setting breakpoints. A breakpoint causes the script to pause, and allows you to enter commands in the Command Window, while the script is paused (it acts just like the keyboard command). Click on line 5 and select Debug ►Set/Clear Breakpoint in the Editor or click:

A red dot appears in a column to the left of line 5. You can also set and clear breakpoints by clicking on the red dots or dashes in this column. To see the line numbers in the Editor, select File ►Preferences, select Editor/Debugger ►Display in that window, and check the Show line numbers option.

In the Command Window, type:

37

```
clear
A = [-1 2 ; 1 -1]
ddom
```

A green arrow appears at line 5, and the prompt K>>
appears in the Command Window. Execution of the script
has paused, just before line 5 is executed. Look at the
variables A and f. Since the diagonal is negative, and f is an
absolute value, we should subtract f from A to preserve the
sign. Type the command:

```
A = A - diag(f)
```

The matrix is now correct, although this works only if all of
the rows need to be fixed and all diagonal entries are
negative. Stop the script by selecting Debug ►Exit Debug
Mode or by clicking:

Alternatively, you can continue where you left off with the
command return or by clicking the Continue icon:

Clear the breakpoint. Replace line 5 with:

```
s = sign(d(i)) ;
A(i,i) = A(i,i) + diag(s .* f(i)) ;
```

Type A=[-1 2;1 -1] and run the script. The script seems
to work, but it modifies A more than is needed. It also fails
on the matrix A=zeros(4) because sign(0) is zero. Fix
the script so that it looks like this:

```
d = diag(A) ;
a = abs(d) ;
f = sum(abs(A), 2) - a ;
```

```
i = find(f >= a) ;
[m n] = size(A) ;
k = i + (i-1)*m ;
tol = 100 * eps ;
s = 2 * (d(i) >= 0) - 1 ;
A(k) = (1+tol) * s .* max(f(i), tol) ;
```

The variable eps (epsilon) gives the smallest value such that
$1+eps > 1$, about 10^{-16} on most computers. It is useful in
specifying tolerances for convergence of iterative processes
and in problems like this one. The odd-looking statement
that computes s is nearly the same as s=sign(d(i)),
except that here we want s to be one when d(i) is zero.

6.3 Function files

Function files provide extensibility to MATLAB. You can
create new functions specific to your problem, which then
have the same status as other MATLAB functions. Variables
in a function file are local by default. A variable can,
however, be declared global (see doc global). Use global
variables with caution; they can be a symptom of bad
program design and can lead to hard-to-debug code.

Convert your ddom.m script into a function by adding these
lines at the beginning of ddom.m:

```
function B = ddom(A)
% B = ddom(A) returns a diagonally
% dominant matrix B by modifying the
% diagonal of A.
```

and add this line at the end of your new function:

```
B = A ;
```

You now have a MATLAB function, with one input
argument and one output argument. To see the difference
between global and local variables as you do this exercise,
type clear. Functions do not modify their inputs, so:

39
```
```

```
C = [1 -2 ; -1 1]
D = ddom(C)
```

returns a matrix D that is diagonally dominant. The matrix C
in the workspace does not change, although a copy of it,
local to the ddom function, called A, is modified as the
function executes. Note that the other variables, a, d, f, i, k
and s no longer appear in your main workspace. Neither do
A and B. These are local to the ddom function.

The first line of the function declares the function name,
input arguments, and output arguments; without this line the
file would be a script file. The MATLAB statement
D=ddom(C) causes the matrix C to be passed as the variable
A in the function and causes the output result to be passed
out to the variable D. Since variables in a function file are
local, their names are independent of those in the current
MATLAB workspace. Your workspace has only the
matrices C and D. If you want to modify C itself, then use
C=ddom(C).

Lines that start with % are comments; more on this in
Section 6.7. An optional return statement causes the
function to finish and return its outputs (this happens
implicitly if execution reaches the end of the M-file). An
M-file can reference other M-files, including itself
recursively.

6.4 Multiple inputs and outputs

A function may also have multiple input and output
arguments. For example, it would be useful to provide the
caller of the ddom function some control over how strong
the diagonal should be and to provide more results, such as
the list of rows (the variable i) that violated diagonal
dominance. Try changing the first line to:

```
function [B,i] = ddom(A, tol)
```

and add a % at the beginning of the line that computes tol. Single assignments can also be made with a function having multiple output arguments. For example, with this version of ddom, the statement D=ddom(C,0.1) assigns the modified matrix to the variable D without returning the vector i. Try it on C=[1 -2 ; -1 1].

6.5 Variable arguments

Not all inputs and outputs of a function need be present when the function is called. The variables nargin and nargout can be queried to determine the number of inputs and outputs present. For example, we could use a default tolerance if tol is not present. Add these statements in place of the line that computed tol:

```
if (nargin == 1)
    tol = 100 * eps ;
end
```

Section 10.1 gives an example of nargin and nargout. Section 7.3 describes the if statement.

Use varargin to create a function that can accept an arbitrary number of inputs. Include a parameter with the exact name varargin as the last input parameter. Any extra arguments are collected in a cell array with the name varargin (see Section 8.1 for a discussion of cell arrays). Try creating this simple function.

```
function x = f(a, b, varargin)
    x = a+b ;
    if (nargin > 2)
        disp (varargin) ;
        c = varargin{1}
    end
end
```

Now try f(2,3) and f(2,3,pi,'whatever'). In the latter case, c is set to pi, the first element in varargin and the third input to f.

Similarly, varargout creates a function with a variable number of outputs. Add varargout as the last output parameter to your function.

```
function [x,varargout] = f(a,b,varargin)
    x = a+b ;
    if (nargin > 2)
        varargout{1} = cos(c) ;
    end
end
```

Now try [x,y]=f(2,3,pi), which computes x=2+3 and y=cos(pi). For a simple yet powerful example of varargin and varargout, see the built-in deal function (enter type deal in the Command Window).

6.6 Unused arguments

Not all output arguments of a function are needed every time the function is used. To ignore arguments that appear at the end of the list of outputs, simply remove them from the list. For example, with your new 2-output ddom function, D=ddom(C) returns only the first argument. The second output i is ignored.

To ignore an argument, use a tilde (~) in its place. For example, to obtain just the indices of the diagonal entries that ddom would modify, use [~,i]=ddom(C). Try it.

6.7 Comments and documentation

The % symbol indicates that the rest of the line is a comment; MATLAB ignores the rest of the line. The first contiguous comment lines are used to document the M-file.

They are available to the online help facility and are displayed if `help ddom` or `doc ddom` are entered. Always include this in your functions. Since you have modified the function to add new inputs and outputs, edit your function to describe the variables `i` and `tol`, and state the default value of `tol`. Next, type `help ddom` or `doc ddom`.

Block comments are useful for lengthy comments or for disabling code that you might want to use later. A block comment starts with a line containing only `%{` and ends with a line containing only `%}`. Block comments in an M-file are not printed by the `help` or `doc` commands.

A line starting with two percent signs (`%%`) denotes the beginning of a MATLAB code cell. This type of cell has nothing to do with cell arrays (discussed in Chapter 8), but defines a section of code in an M-file. Cells can be executed by themselves, and cell publishing (see Chapter 23) creates reports whose sections are defined by the cells of an M-file.

6.8 The MATLAB path

M-files must be in a folder accessible to MATLAB. M-files in the Current Folder, displayed at the top of the main MATLAB window, are always accessible. The current list of directories in the MATLAB search path is obtained by the command `path`. This command can also be used to add or delete directories from the search path. See `doc path`. The command `which` locates functions and files on the path. For example, type `which hilb`. You can modify your MATLAB path with the command `path`, or `pathtool`, which brings up another window.

You can also select File ▶ Set Path. To add a specific folder to your path, right-click a grayed-out folder in the Current Folder window and select Add to Path ▶ Selected Folders.

7 Control Flow Statements

In their basic forms, these MATLAB flow control statements operate like those in most computer languages. Indenting the statements of a loop or conditional statement is optional, but it helps readability to follow a standard convention.

You can type a control flow statement in the Command Window, but this can be difficult to manage. Using an M-file is the easiest way to try these examples.

7.1 The for loop

This loop:

```
n = 10 ; x = []
for i = 1:n
    x = [x, i^2]
end
```

produces a vector of length 10, and

```
n = 10 ; x = []
for i = n:-1:1
    x = [i^2, x]
end
```

produces the same vector. Try them. The vector x grows in size at each iteration. Note that a matrix may be empty (such as x=[ ]). The statements:

```
m = 6 ; n = 4
for i = 1:m
    for j = 1:n
        H(i,j) = 1/(i+j-1) ;
    end
end
H
```

produce and display in the Command Window the 6-by-4 Hilbert matrix. The last H displays the final result. The semicolon on the inner statement is essential to suppress the display of unwanted intermediate results. If you leave off the semicolon, you will see that H grows in size as the computation proceeds. This can be slow if m and n are large. It is more efficient to preallocate the matrix H with the statement H=zeros(m,n) before computing it. Type the command doc hilb and type hilb to see a more efficient way to produce a square Hilbert matrix.

Here is the counterpart of the one-dimensional indexing exercise from pages 29 and 31. It adds 99 to each entry of the matrix that is larger than .5. It is slower than using find or logical indexing.

```
A = rand(5)
[m n] = size(A) ;
for j = 1:n
    for i = 1:m
        if (A(i,j) > .5)
            A(i,j) = A(i,j) + 99 ;
        end
    end
end
A
```

The for statement permits any matrix expression to be used instead of 1:n. The index variable consecutively assumes the value of each column of the expression. For example,

```
s = 0 ;
for c = H
    s = s + sum(c) ;
end
```

computes the sum of all entries of the matrix H by adding its column sums (of course, sum(sum(H)) does it more efficiently; see Section 5.3). Each iteration of the for loop assigns a successive column of H to the variable c. In fact, since 1:n = [1 2 3 ... n], this column-by-column assignment is what occurs with for i = 1:n.

In most statements, the colon operator creates a list of numbers in a MATLAB array. It works differently in a for loop. The assignment i = 1:inf fails because the list is too big to be created. The statement for i=1:inf does not create the list, and becomes just another way of specifying an infinite loop.

7.2   The while loop

The general form of a while loop is:

```
while expression
    statements
end
```

The statements are repeatedly executed as long as the expression remains true. For example, for a given number a, the following computes and displays the smallest nonnegative integer n such that $2^n > a$:

```
a = 1e9
n = 0
while 2^n <= a
    n = n + 1 ;
end
n
```

Note that you can compute the same value n more efficiently by using the log2 function:

```
[f,n] = log2(a)
```

You can terminate a for or while loop with the break statement and skip to the next iteration with the continue statement. Here is an example for both. It prints the odd integers from 1 to 7 by skipping over the even iterations and then terminates the loop when i is 7.

```
for i = 1:10
    if (mod(i,2) == 0)
        continue
    end
    i
    if (i == 7)
        break
    end
end
```

7.3 The if statement

The general form of a simple if statement is:

```
if expression
    statements
end
```

The statements are executed only if the expression is true. Multiple conditions also possible:

```
for n = -2:5
    if n < 0
        parity = 0 ;
    elseif rem(n,2) == 0
        parity = 2 ;
    else
        parity = 1 ;
    end
    disp([n parity])
end
```

The else and elseif are optional. If the else part is used, it must come last. The elseif part is tested only when the if test is false. The else part is executed only when the if test and any elseif tests are all false.

7.4 The switch statement

The switch statement is much like the if statement. If you have one expression that you want to compare against several others, then a switch statement can be more concise than the corresponding if statement. Here is the general format:

```
switch expression
    case expression2
        statements
    otherwise
        statements
end
```

The case statement can be repeated as many times as you want. The otherwise clause is optional. The *expression* in the switch statement must be a scalar or a string. It is compared with each case expression (*expression2*, above), and the statements of the first case that matches are selected and executed. If nothing matches, the otherwise statements are executed. Scalars and strings can be mixed together.

The case expression can be a cell array (discussed in Section 8.1), which is a list of expressions enclosed in curly brackets and separated by commas. The switch expression is compared with each of the expressions in the list, and if any one of them matches, the corresponding statements are executed. Try the example on the next page.

The gallery and why functions have more examples of switch statements. Type edit gallery or edit why to

take a look. See `help switch` or `doc switch` for more
examples.

```
switch x
    case 0
        disp ('x is zero') ;
    case {2, 4, 6, 8}
        disp ('x is 2,4,6, or 8') ;
        x = x / 2 ;
    otherwise
        disp ('x is something else') ;
        x = 0 ;
end
```

7.5 The try/catch statement

Computations can fail for many reasons. MATLAB
variables can morph between matrices, strings, cells, and
other types, and not all MATLAB functions work on all
kinds of types. MATLAB can run out of memory. Files that
you try to read might not exist. These are just a few
examples. It can be very difficult to catch all these cases
before trying an operation that might fail.

If a statement in your MATLAB function fails, you might
want that error to terminate the function. This is what
happens with no `try/catch` block. Alternatively, you
might want to try a computation optimistically, and then
take corrective action if something goes wrong. This is
where the `try/catch` statement is useful. The general
form is:

```
try
    statements
catch ME
    error handling statements
    rethrow (ME)
end
```

The first block of statements is executed. If an error occurs, those statements are terminated, and the second block of statements is executed.

Here is a simple example. These MATLAB statements attempt to load a matrix from a `.mat` file. If the MAT-file is not found, a default matrix is used instead.

```
try
    A = load ('mymatrix.txt') ;
catch
    disp ('could not find mymatrix') ;
    A = magic (5) ;
end
```

The ME variable in the `catch` statement is not required, but it is recommended. This variable keeps a record of what the error is. The name ME is not special, just frequently used (short for MATLAB Exception). If the error-handling code cannot recover, you can use `rethrow(ME)` to reissue the error, which acts as if the error was not caught in the first place.

See `doc try` for more information. More examples of `try/catch` are covered in Sections 9.2 and 14.5.

7.6 Matrix expressions (if and while)

A matrix expression is interpreted by `if` and `while` to be true if every entry of the matrix expression is nonzero. Enter these two matrices:

```
A = [ 1 2 ; 3 4 ]
B = [ 2 3 ; 3 5 ]
```

If you wish to execute a statement when matrices A and B are equal, you could type:

```
if A == B
    disp('A and B are equal')
end
```

If you wish to execute a statement when A and B are not equal, you would type:

```
if any(any(A ~= B))
    disp('A and B are not equal')
end
```

or, more simply,

```
if A == B else
    disp('A and B are not equal')
end
```

Note that the seemingly obvious:

```
if A ~= B
    disp('not what you think')
end
```

does not give what is intended because the statement would execute only if each of the corresponding entries of A and B differ. The functions any and all can be used creatively to reduce matrix expressions to vectors or scalars. Two any's are required above because any is a vector operator (see Section 5.3). In logical terms, any and all correspond to the existential (∃) and universal (∀) quantifiers, respectively, applied to each column of a matrix or each entry of a row or column vector. Like most vector functions, any and all can be applied to dimensions of a matrix other than the columns. An if statement with a two-dimensional matrix expression is equivalent to:

```
if all(all(expression))
    statement
end
```

Two matrices A and B can be checked for equality with
`all(all(A==B))`, but `isequal(A,B)` is simpler. The
`isequal` function works on any kind of variables, such as
strings of different lengths, structs, or cell arrays. Structs
and cell arrays are discussed in the next chapter. Try this
example.

```
isequal('a', 'bb')
all('a' == 'bb')
```

The comparison `'a'=='bb'` results in an error, since the
two strings must have the same length.

7.7 Infinite loops

With loops, it is possible to execute a command that never
stops. Typing Ctrl-C stops a runaway display or
computation. Try:

```
i = 1
while i > 0
    i = i + 1
end
```

then type Ctrl-C to terminate this loop.

8 Advanced Data Structures

Standard arrays are useful data structures for storing lots of
things of the same type, such as an array of numbers or
characters. They cannot be used if you want to store a
number in one position and a string in another. This is
where cell arrays, structs, and objects are useful.
Object-oriented programming is a lengthy topic discussed
in its own chapter (Chapter 9).

8.1 Cell arrays

Cell arrays are collections of other arrays or variables of
varying types and are formed using curly braces. For
example,

```
c = {[3 2 1] ,'I love MATLAB'}
```

creates a cell array. The expression c{1} is a row vector of
length 3, while c{2} is a string. cell(2,3) creates a
2-by-3 array of empty cells (not unlike zeros(2,3)).

A cell array of strings can contain strings of different
lengths, or any variable at all – including another cell array.
Try this:

```
d = {'Why do' ,'I love MATLAB?', c}
```

8.2 Structs

The downside of standard arrays and cell arrays is that each
item is known only by its row and/or column index
(A(1,1), A(2,3), and so on). Use a struct to create a data
structure where each part has its own unique name. Try:

```
clear z
z.particle = 'electron'
z.position = [2 0 3]
z.spin = 'up'
```

The variable z describes a variable with several
characteristics, each with its own type. A cell array can also
include three parts, just like this struct, but they would be
called just z{1}, z{2}, and z{3}.

The clear z statement in the example above makes sure z
does not already exist. If z already exists and is a struct,
then the old struct z is augmented with the new fields,
which might not be what you want.

Cell arrays, structs, and standard arrays can be combined and nested within a struct. You can create an array of structs or a struct containing an array. The next example causes z to become a 2-by-3 array of structs, where z(1,1) is the electron above. z(2,3) is a proton with 'unknown' position and no specified spin. The other entries are empty. Try it.

```
z(2,3).particle = 'proton'
z(2,3).position = 'unknown'
```

Try typing z(1,1), z(2,2), and z(2,3). Every entry in the struct z has the same three fields, but some are empty since they have not been assigned to anything. Note that z(1,1).position and z(2,3).position are not the same data type. Structs are very flexible data structures for representing all kinds of data.

The struct command builds a struct array all at once, using one cell array for each field in the struct. This next one-line statement creates the same struct z. Since it is rather long, try it in your own M-file script, rather than trying to type it in the Command Window.

```
z = struct ( ...
'particle', {'electron',[],[] ;
             [],[],'proton' }, ...
'position', {[2 0 3],    [],[] ;
             [],[],'unknown'}, ...
'spin',     {'up',       [],[] ;
             [],[],[]}) ;
```

Since structs can contain variables of any type, they can contain structs, giving a nested struct. If you type

```
z(1,1).a.b = 42
```

then every entry in the struct array is given a field with the name a, with z(1,1).a containing a scalar b equal to 42.

The a fields of other entries of z are all equal to the empty matrix, [ ]. To access a field defined by a string, use s='a'; z.(s), for example.

Structs are a nice way to keep together all the variables and arrays that define a single problem. For example, in a linear programming problem the goal is to find the vector x that minimizes $c^T x$ where $Ax = b$ and $l \leq x \leq h$, and where A is an m-by-n matrix with $m < n$. These components of the problem are of different sizes, but all can be held in a single struct as (for example) Problem.A, Problem.c, and so on.

8.3 Sets

Sets are represented as vectors, character arrays, or cell arrays of strings. Set operations include:

ismember	test membership, $x \in A$
intersect	set intersection, $A \cap B$
union	set union, $A \cup B$
setdiff	set difference, $A \setminus B$
unique	removes duplicates from a set
setxor	$(A \cup B) \setminus (A \cap B)$
issorted	checks if a set is sorted

Try these examples. Their results are shown to the right.

```
A = [1 2 5 6 8 8 9]
B = [1 2 3 5 7]
ismember(3,A)        % 0
ismember(3,B)        % 1
intersect(A,B)       % [1 2 5]
setdiff(A,B)         % [6 8 9]
union(A,B)           % [1 2 3 5 6 7 8 9]
setxor(A,B)          % [3 6 7 8 9]
unique(A)            % [1 2 5 6 8 9]
```

The inputs to a set operation do not need to be sorted, and they may include duplicate entries. Set operations always return their results with sorted entries and with no duplicates. Note that A includes a duplicate entry, but union(A,B) removes it.

A matrix represents a set with each row being a single element. Character arrays represent sets of strings, all of the same length. For sets of strings with different lengths, use a cell array of strings. A third argument ('rows') must be added if a set (numeric or character) is represented with a matrix instead of a vector.

8.4 Other data types

MATLAB supports many other data types, including logical variables (used for logical indexing in Section 5.8), integers of various sizes, single-precision floating-point variables, sparse matrices, multidimensional arrays, and objects.

The default data type is double, a 64-bit IEEE floating-point number. The single type is a 32-bit IEEE floating-point number which should be used only if you are desperate for memory. A double can represent integers in the range -2^{53} to 2^{53} without any roundoff error, and a double holding an integer value is typically used for loop and array indices. An integer value stored as a double is nicknamed a *flint*. Integer types are typically needed for special cases such as signal processing, image processing, encryption, and bit string manipulation. Integers come in signed and unsigned flavors, and in sizes of 8, 16, 32, and 64 bits. Integer arithmetic is not modular, but saturates on overflow. See doc int8, doc uint8, and doc single for more information.

A sparse matrix is not actually its own data type, but an attribute of the double and logical matrix types. Sparse matrices are stored in a special way that does not require space for zero entries. MATLAB has efficient methods of operating on sparse matrices. Type `doc sparse`, and `doc full`, look in Help:MATLAB ▶User Guide ▶Mathematics ▶Sparse Matrices, or see Chapter 18. Sparse matrices are allowed as arguments for most, but not all, MATLAB operators and functions where a normal matrix is allowed.

`D=zeros(3,5,4,2)` creates a 4-dimensional array of size 3-by-5-by-4-by-2. Multidimensional arrays may also be built up using `cat` (short for concatenation). The first input of `cat` specifies the dimension along which the matrices are to be concatenated, and the next inputs are the matrices to concatenate. Try this example, which creates a 3-by-4-by-2 matrix C.

```
A = ones(3,4)
B = rand(3,4)
C = cat(3, A, B)
```

9 Object-Oriented Programming

Structs are very flexible data structures, but sometimes too flexible. Suppose you want to set `z(1,2).particle` to neutron, but you make a mistake and type this instead:

```
z(1,2).Particle = 'neutron'
```

Oops. You created an entirely new field in the struct, rather than modifying the existing one. Worse yet, every time you pass z to a function, a robust function would need to make sure all the required fields are present. What should it do with fields such as `z.Particle` that it does not recognize? And there is no way to get a list of all functions that can take this struct z as input, or produce one on output.

Enter the MATLAB *object*. Think of it as a specialized struct with fields that cannot change unless you change a

57

file that defines the object (the `classdef`, or class definition file). The `classdef` file specifies everything you can do with or to the object (these are called `methods`). Fields in an object act like the fields of a struct, but you cannot add or remove them arbitrarily.

For this exercise, we will create an object class that represents a matrix factorization. This first version has no methods, just the L and U factors themselves ($LU = A$, where L is lower triangular and U is upper triangular). Open the M-File Editor with a blank `classdef` file (File ▶New ▶Class) and create an M-file called `factor0.m`. The filename and the `classdef` name must match:

```
classdef factor0
    %FACTOR0 my first object
    properties
        L, U ;
    end
end
```

Now in the Command Window, create an object F of class `factor0`, and populate its L and U fields:

```
A = rand(4)
F = factor0
[F.L F.U] = lu(A)
whos
```

Try to create a new field by typing `F.x=0` (for example) in the Command Window. MATLAB refuses to do it, and gives you a warning. Click on the underlined `Methods` in the display of F. This gives a list of the methods that can be used on F. The only method is `factor0`, which creates an empty object with the fields L and U. Clicking on the `factor0` link displayed in the Command Window displays the help for the class.

To add new properties, you must edit the `factor0.m` file. A word of caution: if you edit `factor0.m` when you still

have `factor0` objects in your workspace, you may get a
warning message saying that those objects must be cleared
first before the changes can be applied. Use `clear`
`classes` to clear all your variables and classes. This can
make it difficult to edit and test your code, since you must
clear all your test data. Create another M-file that starts with
`clear classes`, creates some test data, and then tests your
new object class.

The simple `factor0` class acts just like a struct except that
it is not as flexible (which is sometimes a good thing). If G
is not already defined, `[G.L G.U]=lu(A)` acts just like F
in this example, except that G is a struct, not an object of the
`factor0` class. So `G.x=0` works, but `F.x=0` does not.

9.1 Object methods

An object with properties but no methods is useful, but it
usually makes more sense to create a class that contains
`methods` which operate on objects of that class. Create a
file `factor1.m` that contains the following:

```
classdef factor1
    %FACTOR1 my first useful object
    properties
        L, U ;
    end
    methods
        function F = factor1 (A)
            [F.L F.U] = lu (A) ;
        end
        function x = mldivide (F,b)
            x = F.U \ (F.L \ b) ;
        end
    end
end
```

The first method has the same name as the class, and is
called the *constructor*. It creates a `factor1` object, and

must return an object of the class. The constructor for
`factor0` was implicit; it just did `F.L=[]` ; `F.U=[]`.

The second method illustrates an important concept in
object-oriented programming: *overloading*. This
overloaded `mldivide` method defines what happens when
you do x=F\b when F is a `factor1` object. `mldivide` is
the name of the function that does the backslash (\)
operator. Creating an `mldivide` method overloads the \
operator to mean something new for objects of the `factor1`
class.

```
A = rand(4)
b = rand(4,1)
F = factor1(A)
F\b
A\b
```

Both F\b and A\b solve $Ax = b$, but the first one reuses a
previously computed factorization. This is handy, because
you can solve A\c for a new right-hand-side c without
recomputing the LU factorization of A. The `mldivide` (\)
operator cannot do that. Try the example above with a
larger matrix, and use `tic` and `toc` (see Section 10.5) to
compare the performance of x=F\b; and x=A\b;. The
latter takes much more time.

You can of course create new methods which do not
overload existing functions. For example, to create a
method that returns the largest and smallest entries the
diagonal of U, add this function to the `methods` block of
`factor1`:

```
function d = maxdiag (F)
    d = sort (abs (diag (F.U))) ;
    d = d ([1 end]) ;
end
```

The ratio `d(1)/d(2)` is a (very) rough estimate of the condition number of the matrix A whose factorization is held in the object F (see also `cond`, `rcond`, and `condest`).

As an alternative to placing all methods in a single file (as in the `factor1.m` example), you can create a folder name that begins with @, such as `@factor1`. The parent folder of `@factor1` must be on your MATLAB path. Put your `factor1.m` file inside `@factor1`. Remove the `mldivide` and `maxdiag` functions and place them in files called `mldivide.m` and `maxdiag.m` in the `@factor1` folder. The two techniques specify the same class, just with a different file and folder structure. Using an `@folder` is helpful if your methods involve a lot of code.

9.2 Object inheritance and abstract classes

Objects can build on one another to create new objects from old. More precisely, an object class can *inherit* methods and properties from another class. The new class is called a *subclass*, or synonymously a *derived class*, and the class it inherits from is called the *superclass*.

This allows you to write code once (for the superclass) to be used by all the different classes that inherit from the superclass.

This is best illustrated with an example. *LU* factorization is just one kind of factorization. What we would really like is a generic factorization class that can represent any kind of factorization, and select the right one depending on the matrix.

For now, however, let us consider a simpler example which uses just `lu` and `chol` for square dense matrices, and which only provides the A\b function.

The *abstract* class `factor2_generic` specifies a generic object. You cannot create an object of this class. It just provides a framework to write methods and properties used by its subclasses. It defines an `Abstract` method called `mldivide` that must be implemented in any class that inherits from this class. That is done by the two subclasses, `factor2_lu` and `factor2_chol`.

```
classdef factor2_generic
    properties
        L, U ;
    end
    methods (Abstract)
        x = mldivide (F,b) ;
    end
end

classdef factor2_lu < factor2_generic
    %FACTOR2_LU for LU factorization
    methods
        function F = factor2_lu (A)
            [F.L F.U] = lu (A) ;
        end
        function x = mldivide (F,b)
            x = F.U \ (F.L \ b) ;
        end
    end
end

classdef factor2_chol < factor2_generic
    %FACTOR2_CHOL for Cholesky
    methods
        function F = factor2_chol (A)
            F.U = chol (A) ;
        end
        function x = mldivide (F,b)
            x = F.U \ (F.U' \ b) ;
        end
    end
end
```

These two methods inherit all of their properties from `factor2_generic`, using the < syntax in the `classdef` statement. Each of them could also declare additional properties of their own, but they do not need to for this example. They both use the properties from their common superclass, `factor2_generic`, except that `factor2_chol` does not need L, so `F.L` is `[]`.

Using these methods to factorize a matrix requires that you select between the two methods, depending on the matrix. Try this:

```
A = rand(4) ;
C = A'*A ;  % symmetric pos. definite
b = rand(4,1) ;
F = factor2_lu(A) ;
F\b
A\b
G = factor2_chol(C) ;
G\b
C\b
```

Type F and G. Click on the underlined link Superclasses in the display of these two objects in the Command Window, and you will see that both F and G inherit from the `factor2_generic` superclass.

Rather than requiring you to select the right method (`lu` or `chol`), a better approach is to select it automatically with an old-fashioned M-file function (it is not an object-oriented method, but one that returns a `factor2_lu` or `factor2_chol` object). See `factor2.m` on the next page.

Just like the built-in `mldivide`, the `factor2` function tries `chol` if the matrix is symmetric with a real positive diagonal. If that condition does not hold, or if the matrix is not symmetric positive definite, `lu` is used instead. In either case, `F\b` with the resulting F (using `lu` or `chol`) solves $Ax = b$ with the factorization of A. Now you can create F and G without having to choose the method yourself:

```
F = factor2(A)
G = factor2(C)
```

```
function F = factor2 (A)
%FACTOR2 LU or CHOL factorization
chol_ok = false ;
d = diag (A) ;
if (all(d>0) && nnz(imag(d))==0 ...
    && nnz(A-A')== 0)
    try
        F = factor2_chol (A) ;
        chol_ok = true ;
    catch
    end
end
if (~chol_ok)
    F = factor2_lu (A) ;
end
```

9.3 Object attributes

By default, the properties of a class can be read and
modified from methods outside the class, and they are all
displayed by disp(F) if F is an object. That is,
properties act just like the fields of a struct.

This might not be what you want. Try typing F.L='gunk'
with the factor2 object F from the last example. If the
user of the factor2 object has no valid reason to modify a
property, then you should not allow that to happen. This
restriction helps prevent bugs from appearing in your code
that might otherwise go undetected.

There are many property attributes you can modify. The
three most common ones are

- GetAccess, which controls how the property can be
  read,
- SetAccess, which controls how the property can be
  modified, and
- Hidden, which controls whether or not the property
  is displayed by disp or when you leave off the
  semicolon at the end of an assignment statement.

`GetAccess` and `SetAccess` can be `public` (the default), `protected` (access from the class itself or from derived classes), or `private` (access from just the class itself). `Hidden` is either `true` or `false`.

For the `factor2_generic` class, a good choice would be `SetAccess = protected`. This allows subclasses to modify L and U, but it does not allow you to type `F.L='gunk'` at the Command Window, since commands typed there do not appear in the class definition itself. Leave `GetAccess` alone, so that L and U can be extracted. Perhaps you might want to solve x=F.L\b, or extract the diagonal of U with `diag(F.U)`. Those operations make sense, but `F.L='gunk'` does not. If you change `GetAccess = protected`, then x=F.L\b would not be allowed. Which attributes to use depends on how you want the object to be used.

To make this change, modify the `properties` line in `factor2_generic` to look like this:

```
properties (SetAccess = protected)
```

If you want to set more than one attribute, separate them with commas:

```
properties (SetAccess = protected, ...
            GetAccess = public)
```

If you have multiple properties with different attributes, simply include multiple `properties` blocks in the class definition.

With these changes, try `F.L='gunk'` again. It fails gracefully, with a warning that changing the L property is not allowed.

9.4 A more extensive example

A more powerful object for factorizing matrices and solving linear systems can be found in the File Exchange (Section 2.6) by searching for the keyword `factorize`, for the submission entitled

> *Don't let that INV go past your eyes;*
> *To solve that system FACTORIZE.*

The `factorize` object is over 1000 lines long and uses 8 different kinds of factorization methods. It provides methods for A\b, b/A, `disp`, `plus` and `minus` for a low-rank update/downdate, an `inverse` that does not actually compute the inverse, `subsref` for getting entries from A or its inverse, `double` for returning A or inv(A) as a matrix, and `mtimes` for computing inv(A)*b or b*inv(A) without computing inv(A).

It is a useful exercise to go through that lengthy example, since it is a powerful and real-world example of what a MATLAB object can do. It is also posted on the web page for this book.

9.5 Object handle classes

By default, variables in MATLAB have a pass-by-value behavior, including objects. If you pass a variable as an input to a method or to a function, you can change that variable inside the method or function, but the changes do not appear in the caller. Objects that behave like this are from *value* classes.

Handle objects derive from the built-in `handle` class, and behave differently from objects from value classes. Copying an object in the handle class does not create an entirely new instance of the object, but just another reference to the same contents of the object. The new object is just an alias, or another name for the same underlying data. This behavior is useful when the object encapsulates something that cannot be copied, or something you do not want to copy.

For example, suppose an object refers to a person's contact information in a database. If the person's phone number changes, you would like that change to be reflected in all objects that refer to that person. An object that encapsulates the contact information should be a handle class.

Handle classes are also used for MATLAB graphics (Handle Graphics®, discussed in Section 17.1). Copying a figure object does not display a new figure on your screen, but rather gives you another object that refers to the same displayed figure.

For examples of handle classes, see Help:MATLAB ▶User Guide ▶Object-Oriented Programming ▶Value or Handle Class – Which to Use.

10 Advanced M-file Features

This chapter describes advanced M-file techniques, such as how to pass a function as an argument to another function and how to write high-performance code in MATLAB.

10.1 Function handles and anonymous functions

A function handle (@) is a reference to a function that can be treated as a variable. It can be copied, placed in cell array, and evaluated just like a regular function. For example,

```
f = @sqrt
f(2)
sqrt(2)
```

The `str2func` function converts a string to a function handle. For example,

```
f = str2func('sqrt')
f(2)
```

Function handles can refer to built-in MATLAB functions, to your own function in an M-file, or to anonymous functions. An anonymous function is defined with a one-line expression, rather than by an M-file. Try:

```
g = @(x) x^2-5*x+6-sin(9*x)
g(1)
```

Some MATLAB functions that operate on function handles need to evaluate the function on a vector, so it is often better to define an anonymous function (or M-file) so that it can operate entry-wise on scalars, vectors, or matrices. Try this instead:

```
g = @(x) x.^2-5*x+6-sin(9*x)
g([0 1])
```

The general syntax for an anonymous function is

```
handle = @(arg1, arg2, ...) expression
```

Here is an example with two input arguments, which computes the 2-norm of a vector of length 2.

```
norm2 = @(x,y) sqrt(x^2 + y^2)
norm2(4, 5)
norm([4 5])
```

One advantage of anonymous functions is that they can implicitly refer to variables in the workspace or the calling function without having to use the `global` statement. Try this example:

```
A = [3 2 ; 1 3]
b = [3 ; 4]
y = A\b
resid = @(x) A*x-b
resid(y)
A*y-b
```

In this case, x is an argument, but A and b are defined in the calling workspace. To find out what a function handle refers to, use func2str or functions. Try these examples:

```
func2str(f)
func2str(g)
func2str(norm2)
func2str(resid)
functions(f)
```

Cell arrays can contain function handles. They can be indexed and the function evaluated in a single expression. Try this:

```
h{1} = f
h{2} = g
h{1}(2)
f(2)
h{2}(1)
g(1)
```

Here is a more useful example. The bisect function, on the next page, solves the nonlinear equation f(x)=0. It takes a function handle or a string as one of its inputs. If the function is a string, it is converted to a function handle with str2func. bisect also gives you an example of nargin and nargout (see also Section 6.5). Compare bisect with the built-in fzero discussed in Section 21.4.

Type in bisect.m, or download it from the web page for this book.

```
function [b, steps] = bisect(f,x,tol)
% BISECT:  zero of a function of one variable via
% the bisection method.  bisect(f,x) returns a
% zero of the function f.  f is a function handle
% or a string with the name of a function.  x is
% an array of length 2; f(x(1)) and f(x(2)) must
% differ in sign.  An optional third input argument
% sets a tolerance for the relative accuracy of the
% result.  The default is eps.  An optional second
% output argument gives a matrix containing a trace
% of the steps; the rows are of the form [c f(c)].

if (nargin < 3)
    % default tolerance
    tol = eps ;
end
trace = (nargout == 2) ;
if (ischar(f))
    f = str2func(f) ;
end
a = x(1) ;
b = x(2) ;
fa = f(a) ;
fb = f(b) ;
if (trace)
    steps = [a fa ; b fb] ;
end

while (abs(b-a) > 2*tol*max(abs(b),1))
    c = a + (b-a)/2 ;
    fc = f(c) ;
    if (trace)
        steps = [steps ; [c fc]] ;
    end
    if (fb > 0) == (fc > 0)
        b = c ;
        fb = fc ;
    else
        a = c ;
        fa = fc ;
    end
end
```

Try:

```
bisect(@sin, [3 4])
bisect('sin', [3 4])
bisect(g, [0 3])
g(ans)
```

Some MATLAB functions are built in; others are distributed as M-files. The actual listing of any M-file, those in MATLAB or your own, can be viewed with the MATLAB command type. Try entering type eig, type vander, and type rank.

10.2   Name resolution

When MATLAB comes upon a new name, it resolves it into a specific variable or function by checking to see if it is a variable, a built-in function, a file in the current folder, or a file in the MATLAB path (in order of the folders listed in the path). MATLAB uses the first variable, function, or file it encounters with the specified name. You can use the command which to find out what a name is. Try this:

```
clear
which i
i = 3
which i
which i -all
```

10.3   Error and warning messages

Error messages are best displayed with the function error. For example:

```
A = rand(4,3)
[m n] = size(A) ;
if m ~= n
    error('A must be square') ;
end
```

71

This aborts execution of an M-file if the matrix A is not square, and is a useful thing to add to the ddom function that you developed in Chapter 6, since diagonal dominance is only defined for square matrices. Try adding it to ddom (excluding the rand statement, of course), and see what happens if you call ddom with a rectangular matrix.

If you want to print a warning, but continue execution, use the warning statement instead, as in:

```
warning('A singular; computing anyway')
```

See Section 7.5 (try/catch) for one way to recovering from errors. An error statement inside a try block of a try/catch statement causes the catch part to be executed.

10.4   User input

In an M-file the user can be prompted to interactively enter input data, expressions, or commands. When, for example, the statement:

```
iter = input('iteration count: ') ;
```

is encountered, the prompt message is displayed and execution pauses while the user keys in the input data (or, in general, any MATLAB expression). Upon pressing the return or entry key, the data is assigned to the variable iter and execution resumes. You can also input a string; see help input.

An M-file can be paused until a key is typed with the pause command. It is a good idea to display a message, as in:

```
disp('Hit enter to continue: ') ;
pause
```

A Ctrl-C terminates the script or function that is paused. A more general command, keyboard, allows you to type any number of MATLAB commands. See doc keyboard.

10.5 Performance measures

Time and space are the two basic measures of an algorithm's efficiency. In MATLAB, this translates into the number of floating-point operations (flops) performed, the elapsed time, the CPU time, and the memory space used. MATLAB no longer provides a flop count because it uses high-performance block matrix algorithms that make it difficult to count the actual flops performed. On current computers with deep memory hierarchies, flop count is less useful as a performance predictor than it once was.

The elapsed time (in seconds) can be obtained with tic and toc; tic starts the timer and toc returns the elapsed time since the last tic. Hence:

```
tic ; statement ; t = toc
```

returns the elapsed time t for execution of the statement. Type it as one line in the Command Window. Otherwise, the timer records the time you took to type the statement. The elapsed time for solving a linear system above can be obtained, for example, with:

```
n = 1000 ;
A = rand(n) ;
b = rand(n,1) ;
tic ; x = A\b ; t = toc
r = norm(A*x-b)
(2/3)*n^3 / t
```

The norm of the residual is also computed, and the last line reports the approximate flop rate. You may wish to compare x=A\b with x=inv(A)*b for solving the linear system. Try it. You will find A\b to be faster and more accurate.

If there are other programs running at the same time on your computer, elapsed time might not be an accurate measure of performance. Try cputime instead.

MATLAB runs faster if you can restructure your computations to use less time and memory. Here is one practical example. Below are five different ways of applying a Householder transformation to a matrix A, in increasing order of cost. Given a vector x as its second argument, the gallery function for 'house' returns a scalar beta and a column vector v to construct the orthogonal Householder matrix H=eye(m)-beta*v*v', so that H*x is all zero except for the first entry (which is equal to s). Try it.

```
m = 4000 ; n = 1000 ;
A = rand (m,n) ;
[v,beta,s] = gallery('house',A(:,1)) ;
tic ; C = A - v*(beta*(v'*A)) ; toc
tic ; C = A - beta*(v*(v'*A)) ; toc
tic ; C = A - beta*v*v'*A ; toc
tic ; C = A - beta*(v*v')*A ; toc
tic ; H = eye(m)-beta*v*v' ; C=H*A ; toc
```

In practice, H is not formed explicitly. The first two methods do not compute H and require very little time and memory. The third method leaves the order of evaluation for MATLAB to decide, but as of MATLAB 7.10 (R2010a), it computes an m-by-m temporary matrix. The fourth method also computes a matrix of that size (v*v'). The last method explicitly computes the m-by-m matrix H.

Try setting m large enough so that the first two methods succeed but the last three fail by running out of memory or by taking an excessive amount of time. See doc gallery for more information on the 'house' function.

MATLAB does not provide a measure of its peak memory usage, which typically occurs in the middle of a statement, function, or operator. There is no way to determine the memory usage of temporary variables created inside a MATLAB function or operator.

74

Between MATLAB statements, you can find out the total size of your workspace, in bytes, with the command whos.

```
s = whos
space = sum([s.bytes])
```

The memory function (available only on Microsoft Windows) is another option. However, like whos, it only provides memory usage information in between MATLAB statements.

10.6 Efficient code

The ddom function that you wrote in Chapter 6 illustrates some of the MATLAB features that can be used to produce efficient code. All operations are *vectorized*, so that loops are avoided. You could have written the ddom function using nested for loops, much like you would write the function in C, Fortran, or Java. Type in the ddomloops function on the next page, or download it from the web site for this book.

The non-vectorized ddomloops function is only slightly slower than the vectorized ddom (at least for full matrices). In earlier versions of MATLAB, the non-vectorized version would be very slow. MATLAB 6.5 and subsequent versions include a built-in accelerator that greatly improves the performance of non-vectorized code.

Try:

```
A = rand(1000) ;
tic ; B = ddom(A) ; toc
tic ; B = ddomloops(A) ; toc
```

Only simple for loops can be accelerated. Loops that operate on sparse matrices are not accelerated, for example (sparse matrices are discussed in Chapter 18).

```
function B = ddomloops(A,tol)
% B = ddomloops(A) returns a diagonally
% dominant matrix B by modifying the
% diagonal of A.
[m, n] = size(A) ;
if (nargin == 1)
    tol = 100 * eps ;
end
for i = 1:n
    d = A(i,i) ;
    a = abs(d) ;
    f = 0 ;
    for j = 1:n
        if (i ~= j)
            f = f + abs(A(i,j)) ;
        end
    end
    if (f >= a)
        aii = (1 + tol) * max(f, tol) ;
        if (d < 0)
            aii = -aii ;
        end
        A(i,i) = aii ;
    end
end
B = A ;
```

Try:

```
A = sparse(A) ;
tic ; B = ddom(A) ; toc
tic ; B = ddomloops(A) ; toc
```

Since not every loop can be accelerated, writing code that
has no for or while loops is still important. As you
become practiced in writing without loops and reading
loop-free MATLAB code, you will also find that the
loop-free version is often easier to read and understand.

If you cannot vectorize a loop, you can speed it up by
preallocating any vectors or matrices in which output is
stored. For example, by including the second statement

below, which uses the function `zeros`, space for storing E
in memory is preallocated. Without this, MATLAB must
resize E one column larger in each iteration, slowing
execution.

```
M = magic(6) ;
E = zeros(6,50) ;
for j = 1:50
    E(:,j) = eig(M^j) ;
end
```

11   Code Development Tools

The Current Folder window provides a pull-down Actions
menu on its top right corner, which looks like this:

The menu gives you options for creating new M-files of
various types, and options for how files and folders are
viewed in the window (what characteristics are shown, and
how they are sorted and grouped). The Show ►Description
option is particularly useful. It displays the H1 line of each
M-file.

The menu gives you access to six different reports that
provide useful information about your M-files. These tools
are described below in this chapter. Also described is the
File and Folder Comparison tool, which you can find in the
Desktop ►File and Folder Comparisons menu.

11.1   Code Analyzer (M-Lint) report

Navigate to the folder where you created the `ddomloops`
M-file (see Section 10.6), and select ✿▾ ►Reports ►Code
Analyzer Report in the Current Folder window. This report
examines all M-files in the folder and checks them for

suspicious constructs. Scroll down to the report on ddomloops.m, and note that one warning is listed:

5: The value assigned here to 'm' appears to be unused. Consider replacing it with ~.

Click on the underlined 5:. The Editor window opens the ddomloops.m file and highlights the variable m in line 5:

```
[m, n] = size(A) ;
```

The little orange block near the top right of the Editor window tells you the M-Lint Code Analyzer has found problems with your code.

The variable m is assigned by this statement, but not used. This is not an error, just a warning. You could ask MATLAB to fix it automatically. If you hover your mouse over the error and click Fix, and MATLAB will replace m with a tilde (~) to denote an unused output argument. However, this warning should remind you that ddomloops is only intended for square matrices. This is a helpful reminder, because no test is made to ensure the matrix is square. Try:

```
ddomloops(ones(2,3))
```

An obscure error occurs because the non-existent entry A(3,3) is referenced. The function should fail with this input, but the error message is useless.

Save a copy of your original ddomloops.m file and call it ddomloops_orig.m. You will need it for the exercise in Section 11.8.

Add the following code to ddomloops just after line 5:

```
if (m ~= n)
    error('A must be square') ;
end
```

Rerun the Code Analyzer report by clicking this button at the top of the report:

The warning has gone away and your code is more robust. Try ddomloops(ones(2,3)) again. It correctly reports an error that A must be square.

The Code Analyzer can also analyze a single file. In the Editor, select Tools ▶ Show Code Analyzer Report. This option is useful if you have many M-files in a single folder and want to examine just one file at a time. Alternatively, type mlint ddomloops in the Command Window.

11.2   Advanced Editor features

The Editor is a powerful tool for writing, testing, and documenting your M-files. You have already used it to create basic M-file scripts and functions, by working through the examples in Sections 6.1 and 6.2.

Two additional features of the Editor described in this section are *code folding* and *M-file configuration*.

Open one or both of the bisect.m and ddomloops.m files that you worked through earlier in this book (Section 10.1 on page 70, and Section 10.6 on page 76, respectively). Notice the thin lines to the left of the code, with minus signs inside little boxes:

Click one of the boxes. The Editor collapses a region of your code into a single line, such as the body of a for or while loop, or a region of comments. Your code is not gone, of course, just not displayed. This code-folding feature helps you navigate through a large M-file. By

default, cells (Chapter 23), `if/else` blocks, and
`switch/case` blocks cannot be folded, but you can change
this in the File ▶Preferences menu, under the
Editor/Debugger ▶Code Folding tab.

An *M-file configuration* is a way of providing a basic
stand-alone method for testing your M-file functions.
Normally, a function requires inputs. Try clicking the "run"
icon to run one of these functions:

MATLAB complains that the inputs are not defined. The
same thing happens if you type `bisect` or `ddomloops` at
the command line. You can provide a default set of inputs to
your functions by selecting the Debug ▶Run Configuration
for bisect.m menu option (for example). Modify the line
with just the word `bisect`, replacing it with this test case:

```
bisect(@sin, [3 4])
```

Now click the "run" button once more. The `bisect`
function is called with the two inputs you specified in its
M-file configuration, above. This gives you a quick way to
test your M-files, without having to write a separate test
code.

Chapter 23 (*Cell Publishing*) shows you how to use the
Editor to create elegant reports with text, MATLAB code,
results, and figures, which can be recomputed and recreated
with a single push of a button.

11.3 TODO/FIXME report

The TODO/FIXME Report lists all lines in an M-file
containing the words `TODO`, `FIXME`, or `NOTE`, along with the
line numbers in which they appear. Clicking the line
number brings up the Editor at that line. This is useful
during incremental development of a large project.

11.4 Help report

The Help Report examines each M-file in the current folder for the comment lines that appear when you type `help` or `doc` followed by the M-file name. Select ✿· ▶Reports ▶Help Report. Here is the report for `ddomloops`:

```
B = ddomloops(A) returns a diagonally

B = ddomloops(A) returns a diagonally
dominant matrix B by modifying the
diagonal of A.
```

No examples
No See Also line
No copyright line

The first line in the report is the description line, which is the first line after the function statement itself (if the line is a comment line). The MATLAB convention is for the first comment line to be a stand-alone one-line description of the function, starting with the name of the function in all capital letters. This is called the H1 line of an M-file. The first token in this line should be the function name, in capital letters and with no space between the % and the function name.

Edit `ddomloops` and add a new description line, as the second line in the file:

```
%DDOMLOOPS make matrix diagonally dominant
```

Make sure the option ✿· ▶Show ▶Description is checked in the Current Folder window. Take a look at your newly modified `ddomloops.m` function in the window. The helpful one-line H1 description now appears below the filename.

Now go back to the Help Report window. This report complains that there is no example, no *See Also* line, and no copyright line for ddomloops.m. An example starts with a comment line that starts with the word example or Example and ends at the next blank comment line. The *See Also* line is a comment line that starts with the words See also, and is followed by a list of functions related to this function. The copyright line is a comment that starts with the word Copyright, followed by the year and your name. These constructs are all optional, of course, but adding them to the M-file makes the code easier to use. After the last comment line, add the following comments:

```
%
% Example
%     A = [1 0 ; 4 1]
%     B = ddomloops(A)
%     B is the same as A, except B(2,2)
%     is slightly greater than 4.
%
% See also DDOM.
```

Finally, add a blank line (not a comment), and then the line:

```
% Copyright 2010, Me.
```

The function name DDOM appears in upper case, telling MATLAB to recognize it as the name of a function. Rerun the Help Report. All these these constructs are listed in the report. Type help ddomloops or doc ddomloops in the Command Window. Since MATLAB recognizes DDOM as a function name, it underlines the word ddom in blue as an active link. Clicking the link displays the corresponding help or doc for ddom (assuming you created the function when working through the examples in Chapter 6).

Even if you do not plan on sharing your code with someone else, get into the habit of carefully documenting your code

82

with an H1 line and examples. Six months later, when you look at your code again, *you* will be the "someone else" trying to understand your code.

11.5 Contents report

The Contents Report generates a special file called Contents.m that summarizes all of the M-files in a single folder. The Contents.m file is a very useful index when you have lots of M-files in a single folder, particularly if all those files are part of a single large project.

Create a folder entitled diagonal_dominance and place all of the related M-files in this folder (just ddom.m and ddomloops.m, for now). Add the diagonal_dominance folder to your path (see Section 6.8). Now that this folder is on your path, whatever your current folder is, the command help diagonal_dominance lists the Contents.m index, and the ddom and ddomloops functions are always available to you. You can use them in the Command Window, or in other M-files, no matter what your current folder is.

Navigate to this folder in the Current Folder window, and select the Contents Report (❖· ▶Reports ▶Contents Report). Click yes if MATLAB prompts you to create a new Contents.m file, then scroll down until you see your modified ddomloops function. Its name is followed by its one-line description, generated automatically from the description line in ddomloops.m. You can edit Contents.m to add more description, and then click the refresh button to generate a new Contents Report. Any discrepancies are reported to you. For example, if you edit the one-line description in Contents.m, but not in the corresponding M-file, a warning appears and MATLAB offers to fix the discrepancy.

Type the command help diagonal_dominance. This use of the help command prints the Contents.m listing in the diagonal_dominance folder, and highlights the name of each function. Click on ddomloops in the list; the help ddomloops information appears. Many functions in MATLAB are implemented as M-files and are documented in the same way that you have documented your current folder. For example, help general lists the Contents.m file of the folder *MATLAB*/toolbox/matlab/general (where *MATLAB* is the folder in which MATLAB is installed).

11.6 Dependency report

If you would like to share your MATLAB project with someone else, you need to make sure that none of the required M-files in your project are left out. This is where the Dependency Report comes in handy. Assuming that your current folder is diagonal_dominance, select ✿· ▶Reports ▶Dependency Report in the Current Folder window. For each M-file in the current folder, the Dependency Report lists the M-files and mexFunctions that it relies on, and which M-files rely on it. If you see something that your code relies on, but which is *not* in the right place, then you should move it into the diagonal_dominance folder before you share that folder with someone else.

Since ddom and ddomloops are such simple functions, the Dependency Report is not very interesting. Create an M-file script in the diagonal_dominance folder called simple.m:

```
A = [1 2 ; 3 0]
B = ddomloops(A)
C = ddom(A)
```

Re-run the dependency report. `simple` is listed as a parent of its child functions `ddomloops` and `ddom`.

11.7 Profiler tool and Coverage report

MATLAB provides an M-file Profiler that lets you see how much computation time each line of an M-file uses.

Try this example. Create a short M-file script, `ddomtest.m`, ideally in your `diagonal_dominance` folder:

```
A = rand(1000) ;
B = ddomloops(A) ;
```

Then select Tools ▶ Open Profiler in the Editor window, select Desktop ▶ Profiler, or type `profile viewer` in the Command Window. Type `ddomtest` in the text box entitled Run this code and hit enter (or click Start Profiling). A short table appears with the number of calls and time spent in each function. Most of the time is spent in `ddomloops`. Click on the function name and you are given a lengthy description of the time spent in each line of code of `ddomloops`. This report is useful for improving code performance and for debugging. Untested lines of code could harbor a bug.

The Coverage Report provides a short overview of the profile coverage of each file in a folder. Selecting it shows that `ddomtest` was fully exercised (100% coverage), but a few lines of code in `ddomloops` were not tested. The code you added to check for rectangular matrices was not tested, and the case when the diagonal entry `A(i,i)` is negative was not tested.

11.8 File and Folder Comparison tool

The File and Folder Comparisons tool is very useful for tracking changes to your code as you develop it, particularly

for large projects. Select Desktop ▶File and Folder
Comparisons, and click the New File Comparison button.

Select ddomloops_orig.m as one file, and ddomloops.m
as the second file. A color-coded side-by-side display of
these two functions is displayed, showing you the lines that
match and the lines that differ. You may need to adjust the
Columns visible setting to fit both files on your screen. The
up and down arrows to the right of this box navigate to the
next region of code that differs between the two files. The
report should show a single difference, with three lines
appearing in ddomloops.m that do not appear in
ddomloops_orig.m.

You can also select two different folders for comparison, or
two different MAT files.

12 Calling C from MATLAB

There are times when a MATLAB M-file itself is not
enough. You may have a large application or library written
in another language that you would like to use from
MATLAB, or it might be that the performance of your
M-file is not what you would like. Harnessing the strengths
of both C and M is a powerful problem-solving combination
(see page 232).

MATLAB can call routines written in C, Fortran, or Java.
Similarly, programs written in C and Fortran can call
MATLAB. In this chapter, we will just look at how to call a
C routine from MATLAB. For more information, see
Help:MATLAB ▶User Guide ▶External Interfaces, or see the
online PDF documents *External Interfaces* and *C and
Fortran API Reference*. The discussion in this chapter
assumes that you already know C.

12.1 A simple example

A routine written in C that can be called from MATLAB is
called a MEX-file. The routine must always have the name
mexFunction, and the arguments to this routine are always
the same, regardless of what the mexFunction does. Here is
a very simple MEX-file; type it in as the file hello.c in the
MATLAB Editor or in your favorite text editor.

```c
#include "mex.h"
void mexFunction
(
    int nargout,
    mxArray *pargout [ ],
    int nargin,
    const mxArray *pargin [ ]
)
{
    mexPrintf ("hello world\n") ;
}
```

Compile and run it by typing:

```
mex hello.c
hello
```

If this is the first time you have compiled a C MEX-file on a
PC with Microsoft Windows, you will be prompted to select
a C compiler.

The arguments nargout and nargin are the number of
outputs and inputs to the function (just as an M-file
function), and pargout and pargin are pointers to the
arguments themselves (of type mxArray).

The mexPrintf function is just the same as printf in C.
You can also use printf itself; the mex command redefines
it as mexPrintf with a #define when the program is
compiled. This way, you can write a routine that can be
used from MATLAB or from a stand-alone C application,
without MATLAB.

Although this `hello.c` MEX-file does not have any inputs or outputs, you could modify it to print out `nargin` and `nargout`, and try this:

```
[a,b] = hello('there')
```

MATLAB complains that the two outputs are not assigned, but this is just a starter example.

12.2 C versus MATLAB arrays

MATLAB stores its arrays in column major order, while the convention for C is to store them in row major order. Also, the number of columns in an array is not known until the mexFunction is called. Thus, two-dimensional arrays in MATLAB must be accessed with one-dimensional indexing in C (see also Section 5.7). In the example in the next section, the `INDEX` macro helps with this translation.

Array indices also appear differently. MATLAB is written primarily in C/C++, and it stores all of its arrays internally using zero-based indexing in which an m-by-n matrix has rows 0 to m-1 and columns 0 to n-1. However, when accessed via an M-file or via commands in the Command Window, MATLAB variables use one-based indexing, where an m-by-n matrix has rows 1 to m and columns 1 to n. When you type the MATLAB command x=A(i,j), MATLAB subtracts one from i and j to reflect this difference. Likewise, indices returned by mexFunctions must start with one, not zero. In the example below, one is added to the `List` array returned by `diagdom` to account for this difference. As an author of C mexFunctions, you will need to constantly translate between zero-based and one-based indexing.

12.3 A matrix computation in C

In Chapters 6 and 10, you wrote the function ddom.m. Here is the same function written as an ANSI C MEX-file. Download the code from the web site for the book, or type in these three files: diagdom.c, diagdom.h, and diagdom_mex.c.

Below is the diagdom.h definitions file. You should use either ptrdiff_t or mwSignedIndex as the default integer type inside a MATLAB mexFunction. This type is 32 bits in size on 32-bit computers, and 64 bits on a 64-bit computer. If you are using a 32-bit version of MATLAB on a 64-bit computer, then change the #define to mwSignedIndex.

```
#include <float.h>
#include <stddef.h>
#define INDEX(i,j,m) ((i)+(j)*(m))
#define ABS(x) ((x) >= 0 ? (x) : -(x))
#define MAX(x,y) (((x)>(y)) ? (x):(y))
#define INT ptrdiff_t
```

The main computational routine, diagdom.c, is on the next page. Compare it with with the loop-based version ddomloops.m in Section 10.6. The logic is very similar, since C has no vectorized expressions. The diagdom_mex.c file is the gateway routine that connects MATLAB with your computational routine. Appending _mex to this filename is not required; it is just a useful practice so that you can easily tell the two files apart. MATLAB mx and mex routines are described in Section 12.4. Place the files in your diagonal_dominance folder, and try this example:

```
mex diagdom.c diagom_mex.c
A = rand(6) ;
B = ddom(A) ;
C = diagdom(A) ;
```

```c
#include "diagdom.h"
void diagdom
(
    double *A, INT n, double *B,
    double tol, INT *List, INT *nList
)
{
    double d, a, f, bij, bii ;
    INT i, j, k ;
    for (k = 0 ; k < n*n ; k++)
    {
        B [k] = A [k] ;
    }
    if (tol < 0)
    {
        tol = 100 * DBL_EPSILON ;
    }
    k = 0 ;
    for (i = 0 ; i < n ; i++)
    {
        d = B [INDEX (i,i,n)] ;
        a = ABS (d) ;
        f = 0 ;
        for (j = 0 ; j < n ; j++)
        {
            if (i != j)
            {
                bij = B [INDEX (i,j,n)] ;
                f += ABS (bij) ;
            }
        }
        if (f >= a)
        {
            List [k++] = i ;
            bii = (1 + tol) * MAX (f, tol) ;
            if (d < 0)
            {
                bii = -bii ;
            }
            B [INDEX (i,i,n)] = bii ;
        }
    }
    *nList = k ;
}
```

The diagdom_mex.c file is listed on the next two pages.

90

```c
#include "mex.h"
#include "diagdom.h"

void error (char *s)
{
    mexPrintf
    ("Usage: [B,i] = diagdom (A,tol)\n") ;
    mexErrMsgTxt (s) ;
}

void mexFunction
(
    int nargout, mxArray *pargout [ ],
    int nargin,  const mxArray *pargin [ ]
)
{
    double tol, *A, *B, *I ;
    INT n, k, *List, nList ;

    /* get inputs A and tol */
    if (nargout > 2 || nargin > 2 || nargin==0)
    {
        error ("Wrong number of arguments") ;
    }
    if (mxIsSparse (pargin [0]))
    {
        error ("A cannot be sparse") ;
    }
    n = mxGetN (pargin [0]) ;
    if (n != mxGetM (pargin [0]))
    {
        error ("A must be square") ;
    }
    A = mxGetPr (pargin [0]) ;
    tol = -1 ;
    if (nargin > 1)
    {
        if (!mxIsEmpty (pargin [1]) &&
            mxIsDouble (pargin [1]) &&
            !mxIsComplex (pargin [1]) &&
            mxIsScalar (pargin [1]))
        {
            tol = mxGetScalar (pargin [1]) ;
        }
        else
        {
            error ("tol must be scalar") ;
        }
```

```
    }

    /* create output B */
    pargout [0] =
        mxCreateDoubleMatrix (n, n, mxREAL) ;
    B = mxGetPr (pargout [0]) ;

    /* get temporary workspace */
    List = (INT *) mxMalloc (n * sizeof (INT)) ;

    /* do the computation */
    diagdom (A, n, B, tol, List, &nList) ;

    /* create output I */
    pargout [1] =
        mxCreateDoubleMatrix (nList, 1, mxREAL);
    I = mxGetPr (pargout [1]) ;
    for (k = 0 ; k < nList ; k++)
    {
        I [k] = (double) (List[k] + 1) ;
    }

    /* free the workspace */
    mxFree (List) ;
}
```

The matrices B and C are the same (round-off error might cause them to differ slightly). On my MacBook Air, the C mexFunction diagdom is about twice as fast than the M-file ddom for large matrices. However, ddom.m is much easier to write and maintain. The performance gap between C and MATLAB continually drops in newer versions of MATLAB, so switching to a mexFunction is less and less a requirement. Use a mexFunction only if the programming effort is worth the payoff.

The Profiler (Section 11.7) comes in handy for finding performance bottlenecks in your code. If you are unable to optimize your M-file to a sufficient degree, you can try replacing select bottlenecks with mexFunctions. Be advised that for some operations, the built-in functions can be many times faster than a simple C mexFunction. For example, x=A\b for large and/or sparse matrices can be an orders of magnitude faster than a simple C (or Fortran) mexFunction for computing the same thing.

12.4 MATLAB mx and mex routines

In the last example, the C gateway routine (in diagdom_mex.c) calls several MATLAB routines with the prefix mx or mex. Routines with mx prefixes operate on MATLAB matrices and include:

mxIsEmpty	1 if the matrix is empty, 0 otherwise
mxIsSparse	1 if the matrix is sparse, 0 otherwise
mxGetN	number of columns of a matrix
mxGetM	number of rows of a matrix
mxGetPr	pointer to the real values of a matrix
mxGetScalar	the value of a scalar
mxCreateDoubleMatrix	create MATLAB matrix
mxMalloc	like malloc in ANSI C
mxFree	like free in ANSI C

Routines with mex prefixes operate on the MATLAB environment and include:

mexPrintf	like printf in C
mexErrMsgTxt	like MATLAB error statement
mexFunction	the gateway routine from MATLAB

Many other mx and mex routines are available. The memory management routines in MATLAB (mxMalloc, mxFree, and mxCalloc) are much easier to use than their ANSI C counterparts. If a memory allocation request fails, the mexFunction terminates and control is passed backed to MATLAB. Any workspace allocated by mxMalloc that is not freed when the mexFunction returns or terminates is automatically freed by MATLAB. This is why no memory allocation error checking is included in diagdom_mex.c; it is not necessary.

Note that all of the references to MATLAB mx and mex routines are limited to the mexFunction gateway routine (diagdom_mex.c). This is not required; it is just a good idea, because you can use your diagdom.c function in other codes, not just in a MATLAB mexFunction.

In this example, no memory allocation was required in diagdom.c itself. If you need to allocate memory, using the C malloc function is dangerous because it can lead to memory leaks. MATLAB does not know about memory allocated by malloc, but just mxMalloc instead.

This leads to a conundrum. How do you avoid using mx functions in your computational routine, while at the same time use mxMalloc? The solution is to exploit the MATLAB_MEX_FILE macro, which is only defined when compiling with the MATLAB mex command. If you add the following code to diagdom.h, then you can use MALLOC and FREE instead of the C or MATLAB versions of these functions.

```
#ifdef MATLAB_MEX_FILE
#include "mex.h"
#define MALLOC mxMalloc
#define FREE mxFree
#else
#include <stdlib.h>
#define MALLOC malloc
#define FREE free
#endif
```

Alternatively, if you have an existing set of files in C and do not want to replace malloc with the MALLOC macro throughout your code, try this, which replaces malloc with a macro.

```
#ifdef MATLAB_MEX_FILE
#include "mex.h"
#define malloc mxMalloc
#define free mxFree
#endif
```

You can do the same with calloc and realloc, if you need the MATLAB versions of those standard C functions.

12.5 Online help for MEX routines

Create an M-file called `diagdom.m` and place it in your `diagonal_dominance` folder:

```
function [B,i] = diagdom(A,tol)
%DIAGDOM:  modify the matrix A.
% [B,i] = diagdom(A,tol) returns a
% diagonally dominant matrix B by
% modifying the diagonal of A.  i is a
% list of modified diagonal entries.
error('diagdom mexFunction not found');
```

Now type `help diagdom` or `doc diagdom`. This is a simple method for providing online help for your own MEX-files.

If both `diagdom.m` and the compiled `diagdom` mexFunction are on the MATLAB path, then the `diagdom` mexFunction is called. If only the M-file is in the path, it is called instead; thus the error statement in `diagdom.m` above. This usually means that the `diagdom` mexFunction has not yet been compiled. Triggering an error in this M-file is much better than returning silently.

12.6 Larger examples on the web

You can find many C/C++ mexFunctions in the File Exchange at MATLAB Central (see Section 2.6). The site www.cise.ufl.edu/research/sparse contains many mexFunctions for sparse matrix computations (see Chapter 18).

13 Calling Fortran from MATLAB

C and C++ are great languages for numerical calculations, particularly if the data structures are complicated. MATLAB itself is written primarily in C/C++, except for many of the core dense matrix computations (LAPACK and the BLAS), which are written in Fortran. In this chapter we will look at how to call a Fortran subroutine from MATLAB. A Fortran subroutine is accessed via a mexFunction in much the same way as a C subroutine is called. Normally, the mexFunction acts as a gateway routine that gets its input arguments from MATLAB, calls a computational routine, and then returns its output arguments to MATLAB, just like the C example in the previous chapter.

13.1 Solving a transposed system

The linsolve function was introduced in Section 5.5, with an example that solves x=U'\b where U is dense, square, real, and upper triangular. Here is a computational routine written in Fortran that computes the same thing. It has no calls to MATLAB-specific mx or mex routines. Both linsolve (with the right opts input) and utsolve solve the system x=U'\b without explicitly forming the transpose, U'.

```
      subroutine utsolve (n, x, A, b)
      integer n
      real*8 x(n), A(n,n), b(n), xi
      integer i, j
      do 1 i = 1,n
        xi = b(i)
        do 2 j = 1,i-1
          xi = xi - A(j,i) * x(j)
2       continue
        x(i) = xi / A(i,i)
1     continue
      return
      end
```

13.2 A Fortran mexFunction with %val

To call this computational subroutine from MATLAB as
x=utsolve(A,b), we need a gateway routine, the first
lines of which must be:

```
      subroutine mexFunction
     $ (nargout, pargout, nargin, pargin)
      integer nargout, nargin
      integer pargout (*), pargin (*)
```

where the $ is in column 6. These lines must be the same
for any Fortran mexFunction (you do not need to split the
first line). Note that pargin and pargout are arrays of
integers. If you have a 64-bit version of MATLAB, use
integer*8 for all integers except nargin and nargout.

MATLAB passes its inputs and outputs as pointers to
objects of type mxArray, but Fortran cannot handle
pointers. Most Fortran compilers can convert integer
"pointers" to references to Fortran arrays via the
non-standard %val construct. We will use this in our
gateway routine. The next two lines of the gateway routine
declare some MATLAB mx routines.

```
      integer mxGetN, mxGetPr
      integer mxCreateDoubleMatrix
```

This is required because standard Fortran has no include-file
mechanism. The next lines determine the size of the input
matrix and create the n-by-1 output vector x. The variable
one is needed if you are on a 64-bit MATLAB and use
integer*8 variables.

```
      integer n, one
      one = 1
      n = mxGetN (pargin (1))
      pargout (1) =
     $ mxCreateDoubleMatrix (n, one, 0)
```

We can now convert "pointers" into Fortran array references and call the computational routine.

```
   call utsolve (n,
$ %val (mxGetPr (pargout (1))),
$ %val (mxGetPr (pargin (1))),
$ %val (mxGetPr (pargin (2))))
   return
   end
```

The arrays in both MATLAB and Fortran are column-oriented and one-based, so translation is not necessary as it was in the C mexFunction.

Combine the two routines into a single file called utsolve.f and type:

```
   mex utsolve.f
```

in the MATLAB Command Window. Error checking could be added to ensure that the two input arguments are of the correct size and type. The code would look much like the C example in Chapter 12, so it is not included. Test this routine on as large a matrix that your computer can handle.

```
n = 5000
A = triu(rand(n,n)) ;
b = rand(n,1) ;
tic ; x = A'\b ; toc
opts.UT = true
opts.TRANSA = true
tic ; x2 = linsolve(A,b,opts) ; toc
tic ; x3 = utsolve(A,b) ; toc
norm(x-x2)
norm(x-x3)
```

The solutions should agree quite closely. In older versions of MATLAB (7.6 and earlier), linsolve and utsolve are an order of magnitude faster than x=A'\b. They require less memory, as well, since they do not have to construct A′. Recent versions (7.9 and later) treat ′\ as a single operator, so that the transpose is not formed when computing x=A′\b. In these versions of MATLAB, utsolve, linsolve, and x=A′\b have about the same performance. linsolve was introduced in MATLAB 7.0.

13.3 If you cannot use %val

If your Fortran compiler does not support the %val construct, then you will need to call MATLAB mx routines to copy the MATLAB arrays into Fortran arrays, and vice-versa. The GNU f77 compiler supports %val, but issues a warning that you can safely ignore. In this utsolve example, add this to your mexFunction gateway routine:

```
integer nmax
parameter (nmax = 5000)
real*8 A(nmax,nmax), x(nmax), b(nmax)
```

where nmax is the largest dimension you want your function to handle. Unless you want to live dangerously, you should check n to make sure it is not too big:

```
if (n .gt. nmax) then
    call mexErrMsgTxt ("n too big")
endif
```

Replace the call to utsolve with this code.

```
    call mxCopyPtrToReal8
$ (mxGetPr (pargin (1)), A, n**2)
    call mxCopyPtrToReal8
$ (mxGetPr (pargin (2)), b, n)
    call utsolve (n, x, A, b)
    call mxCopyReal8ToPtr
$ (x, mxGetPr (pargout (1)), n)
```

This copies the input MATLAB arrays A and b to their Fortran counterparts, calls the `utsolve` routine, and then copies the solution x to its MATLAB counterpart. Although this is more portable, it takes more memory and is significantly slower. If possible, use `%val`.

14 Calling Java from MATLAB

While C and Fortran excel at numerical computations, Java is well-suited to web-related applications and graphical user interfaces. MATLAB can handle native Java objects in its workspace and can directly call Java methods on those objects. No mexFunction is required.

14.1 A simple example

Try this in the MATLAB Command Window:

```
t = 'hello world'
s = java.lang.String(t)
s.indexOf('w') + 1
find(s == 'w')
whos
```

You have just created a Java string in the MATLAB workspace, and determined that the character `'w'` appears as the seventh entry in the string using both the `indexOf` Java method and the `find` MATLAB function.

14.2 Encryption/decryption

MATLAB can handle strings on its own, without help from Java, of course. Here is a more interesting example. Type in the following as the M-file `getkey.m` or download it from the web site for this book.

```
function key = getkey(password)
%GETKEY: key = getkey(password)
% Converts a string into a key for use
% in the encrypt and decrypt functions.
% Uses triple DES.
import javax.crypto.spec.*
b = int8(password) ;
n = length(b) ;
b((n+1):24) = 0 ;
b = b(1:24) ;
key = SecretKeySpec(b, 'DESede') ;
```

The getkey routine takes a password string and converts it into a 24-byte triple DES key using the javax.crypto package. You can then encrypt a string with the encrypt function:

```
function e = encrypt(t, key)
%ENCRYPT: e = encrypt(t, key)
% Encrypt the plaintext string t into
% the encrypted byte array e using a key
% from getkey.
import javax.crypto.*
cipher = Cipher.getInstance('DESede') ;
cipher.init(Cipher.ENCRYPT_MODE, key) ;
e = cipher.doFinal(int8(t))' ;
```

Except for the function statement and the comments, this looks almost exactly the same as the equivalent Java code. This is not a Java program, however, but a MATLAB M-file that uses Java objects and methods. Finally, the decrypt function undoes the encryption:

```
function t = decrypt(e, key)
%DECRYPT: t = decrypt(e, key)
% Decrypt the encrypted byte array e
% into to plaintext string t using a key
```

```
% from getkey.
import javax.crypto.*
cipher = Cipher.getInstance('DESede') ;
cipher.init(Cipher.DECRYPT_MODE, key) ;
t = char(cipher.doFinal(e))' ;
```

With these three functions in place, try:

```
k = getkey('this is a secret password')
e = encrypt('a hidden message',k)
decrypt(e,k)
```

Now you encrypt and decrypt strings in MATLAB (a
feature MATLAB does not provide on its own).

14.3   Java class path

If you define your own Java classes that you want to use
within MATLAB, you need to modify your Java class path.
This path is different than the path used to find M-files. You
can add folders to the static Java path by editing the file
classpath.txt, or you can add them to your dynamic
Java path with the command

```
javaaddpath folder
```

where *folder* is the name of a folder containing compiled
Java classes. javaclasspath lists the folders where
MATLAB looks for Java classes. Use the command which
classpath.txt to find where your static class path file is
located. If you do not have write permission to
classpath.txt, you need to type the javaaddpath
command every time you start MATLAB. It is easier to do
this automatically by creating an M-file script called
startup.m and placing in it the javaaddpath command.
Place your startup.m file in one of the folders in your
MATLAB path, or in the folder in which you launch
MATLAB, and it will be executed whenever MATLAB
starts.

14.4 Calling your own Java methods

To write your own Java classes that you can call from MATLAB, you must first download and install the Java SDK (Software Development Kit) from java.sun.com. You may need to edit your operating system's PATH variable so that you can type the command javac in your operating system command prompt.

MATLAB includes two M-files that can download a web page into either a string (urlread) or a file (urlwrite). Try:

```
s = urlread('http://www.mathworks.com')
```

The urlread function is an M-file. You can take a look at it with the command edit urlread. It uses a Java package from The MathWorks called mlwidgets.io.InterruptibleStreamCopier, but only the compiled class file is distributed, not the Java source file. Create your own URL reader, a purely Java program, and put it in a file called myreader.java. The code appears on the next page.

The geturl method opens the URL given by the string u, and copies it into a file whose name is given by the string f. You can compile this Java program and run it by typing these commands at your operating system command prompt:

```
javac myreader.java
java myreader http://www.google.com my.txt
```

The second command copies Google's home page into your own file called my.txt. You can also type the commands in the MATLAB Command Window, as in:

```
!javac myreader.java
```

```java
import java.io.* ;
import java.net.* ;
public class myreader
{
  public static void main (String [ ] args)
  {
    geturl (args [0], args [1]) ;
  }
  public static void geturl (String u, String f)
  {
    try
    {
      URL url = new URL (u) ;
      InputStream i = url.openStream ();
      OutputStream o = new FileOutputStream (f);
      byte [ ] s = new byte [4096] ;
      int b ;
      while ((b = i.read (s)) != -1)
      {
        o.write (s, 0, b) ;
      }
      i.close ( ) ;
      o.close ( ) ;
    }
    catch (Exception e)
    {
      System.out.println (e) ;
    }
  }
}
```

Now that you have your own Java method, you can call it from MATLAB just as the java.lang.String and javax.crypto.* methods. In the MATLAB Command Window, type (as one line):

```
myreader.geturl
('http://www.google.com','my.txt')
```

14.5   Loading a URL as a matrix

An even more interesting use of the myreader.geturl method is to load a MAT-file or ASCII file from a web page directly into the MATLAB workspace as a matrix. Here is a

simple `loadurl` M-file that does just that. It can read
compressed files; the Java method uncompresses the URL
automatically if it is compressed.

```
function result = loadurl(url)
% result = loadurl(url)
% Reads the URL given by the input
% string, url, into a temporary file
% using myread.java, loads it into a
% MATLAB variable, and returns the
% result. The URL can contain a MATLAB-
% readable text file, or a MAT-file.
t = tempname ;
myreader.geturl(url, t) ;
try
    result = load(t) ;
catch
    try
        result = load('-mat', t) ;
    catch
        result = [ ] ;
    end
end
if (exist(t, 'file'))
    delete(t) ;
end
```

Try it with a simple text file (type this in as one line):

```
w = loadurl('http://www.cise.ufl.edu/
~davis/MATLABPrimer8E/w')
```

which loads in a 2-by-2 matrix. Also try it with this rather
lengthy URL (type the string on one line). `spy` plots a
sparse matrix (see Section 18.5).

```
s = loadurl('http://www.cise.ufl.edu/
research/sparse/mat/HB/west0479.mat')
prob = s.Problem
spy(prob.A)
title([prob.name ': ' prob.title])
```

15 Two-Dimensional Graphics

MATLAB is a powerful tool for creating beautiful plots with very little effort on your part. The primary command for creating two-dimensional plots is plot. Chapter 16 discusses three-dimensional graphics. To preview some of these capabilities, enter the command demo and select some of the visualization and graphics demos. See Chapter 19 for a discussion of how to plot symbolic functions.

15.1 Planar plots

The plot command creates linear *x-y* plots; if x and y are vectors of the same length, the command plot(x,y) opens a graphics window and draws an *x-y* plot of the elements of y versus the elements of x. You can, for example, draw the graph of the sine function over the interval -4 to 4 with the following commands:

```
x = -4:0.01:4 ;
y = sin(x) ;
plot(x, y) ;
```

Try it. The vector x is a partition of the domain with mesh size 0.01, and y is a vector giving the values of sine at the nodes of this partition (recall that sin operates entry-wise). When plotting a curve, the plot routine is actually connecting consecutive points induced by the partition with line segments. Thus, the mesh size should be chosen sufficiently small so that the curve is smooth.

The next example draws the graph of $y = e^{-x^2}$ over the interval -3 to 3. Note that you must precede ^ by a period to ensure that it operates entry-wise:

```
x = -3:.01:3 ;
y = exp(-x.^2) ;
plot(x, y) ;
```

Select Tools ►Zoom In or Tools ►Zoom Out in the Figure window to zoom in or out, or click these buttons (or see the zoom command):

Then click on the figure where you want to zoom in or out. The Tools ►Pan option allows you to click and drag the range and domain displayed by the Figure window.

15.2 Multiple figures

You can have several concurrent Figure windows, one of which will at any time be the designated current figure in which graphs from subsequent plotting commands will be placed. If, for example, Figure 1 is the current figure, then the command figure(2) (or simply figure) opens a second figure (if necessary) and makes it the current figure. The command figure(1) exposes Figure 1 and makes it again the current figure. The command gcf returns the current figure number, and figure(gcf) brings the current figure window to the front.

Just like any other window, a Figure window can be docked in the main MATLAB window. If you have many figures, try using the Figures window. It is available as a Desktop tool under the Desktop ►Figures menu item, and it also comes up automatically whenever you dock a Figure. Without this tool, all of your figures appear by default as separate undocked windows in MATLAB. If you have a lot of figures, your desktop can become hard to manage with many cluttered windows.

MATLAB does not draw a plot right away. It waits until all computations are finished, until a figure command is encountered, or until the script or function requests user input (see Section 10.4). To force MATLAB to draw a plot right away, use the command `drawnow`. This command is very useful to animate the results of your computations as they are computed.

15.3 Graph of a function

MATLAB supplies a function `fplot` to plot the graph of a function. For example, to plot the graph of the function in the last example, you can first define the function in an M-file called, say, `expnormal.m` containing:

```
function y = expnormal(x)
y = exp(-x.^2) ;
```

Then:

```
fplot(@expnormal, [-3 3])
```

produces the graph over the indicated *x*-domain. Using an anonymous function gives the same result without creating `expnormal.m`:

```
f = @(x) exp(-x.^2)
fplot(f, [-3 3])
```

15.4 Parametrically defined curves

Plots of parametrically defined curves can also be made:

```
t = 0:.001:2*pi ;
x = cos(3*t) ;
y = sin(2*t) ;
plot(x, y) ;
```

15.5 Titles, labels, text in a graph

The graphs can be given titles, axes can be labeled, and text can be placed within the graph with the following commands, which take a string as an argument.

title	graph title
xlabel	x-axis label
ylabel	y-axis label
gtext	place text on graph using the mouse
text	position text at specified coordinates

For example, the command:

```
title('A parametric cos/sin curve')
```

gives a graph a title. The command gtext('The Spot') lets you interactively place the designated text on the current graph by placing the mouse crosshair at the desired position and clicking the mouse. It is a good idea to prompt the user before using gtext. To place text in a graph at designated coordinates, use the command text (see doc text). These commands are also in the Insert menu in the Figure window. Select Insert ►TextBox, click on the figure, type something, and then click somewhere else to finish entering the text. If the Edit Plot button

is depressed (or select Tools ►Edit Plot), you can right-click on anything in the figure and see a pop-up menu that gives you options to modify the item you just clicked. You can click and drag objects on the figure. Selecting Edit ►Axes Properties brings up a window with many more options. For example, clicking the Grid: X, Y boxes adds grid lines (as does the grid command).

15.6 Control of axes and scaling

By default, MATLAB scales the axes itself (auto-scaling). This can be overridden by the command `axis` or by selecting Edit ►Axes Properties. Some features of the axis command are:

```
axis([xmin xmax ymin ymax]) sets the axes
```

axis manual	freezes the current axes for new plots
axis auto	returns to auto-scaling
v = axis	vector v shows current scaling
axis square	axes of same size (not same scale)
axis equal	same scale and tic marks on axes
axis off	removes the axes
axis on	restores the axes

The `axis` command should be given after the `plot` command. Try `axis([-2 2 -3 3])` with the current figure. Note that text entered on the figure using the `text` or `gtext` commands moves as the scaling changes (think of it as attached to the data you plotted). Text entered via Insert ►TextBox stays put.

15.7 Multiple plots

Here is one way to make multiple plots on a single graph:

```
x = 0:.01:2*pi;
y1 = sin(x) ;
y2 = sin(2*x) ;
y3 = sin(4*x) ;
plot(x, y1, x, y2, x, y3)
```

Another method uses a matrix y containing the functional values as columns:

```
x = (0:.01:2*pi)' ;
y = [sin(x), sin(2*x), sin(4*x)] ;
plot(x, y)
```

The x and y vectors must have the same length, but each pair can have different lengths. Try:

```
plot(x, y, [0 2*pi], [0 0])
```

The command hold on freezes the current graphics screen so that subsequent plots are superimposed on it. The axes may, however, become rescaled. Entering hold off releases the hold. clf clears the figure, and close closes it. legend places a legend in the current figure to identify the different graphs. See doc legend.

15.8 Line types, marker types, colors

You can override the default line types, marker types, and colors. For example,

```
x = 0:.01:2*pi ;
y1 = sin(x) ;
y2 = sin(2*x) ;
y3 = sin(4*x) ;
plot(x,y1, '--', x,y2, ':', x,y3, 'o')
```

renders a dashed line and dotted line for the first two graphs, whereas for the third the symbol o is placed at each node. The line types are:

'-'	solid	':'	dotted
'--'	dashed	'-.'	dashdot

and the marker types are:

'.'	point	'o'	circle
'x'	x-mark	'+'	plus
'*'	star	's'	square
'd'	diamond	'v'	triangle-down
'^'	triangle-up	'<'	triangle-left
'>'	triangle-right	'p'	pentagram
'h'	hexagram		

Colors can be specified for the line and marker types:

'y'	yellow	'm'	magenta
'c'	cyan	'r'	red
'g'	green	'b'	blue
'w'	white	'k'	black

Thus, plot(x,y1,'r--') plots a red dashed line.

15.9 Subplots and specialized plots

The command subplot(m,n,p) partitions a single figure into an m-by-n array of panes, and makes pane p the current plot. The panes are numbered left to right. A subplot can span multiple panes by specifying a vector p. Here is the last example with each data set plotted in a separate subplot:

```
subplot(2,2,1)
plot(x,y1, '--')
subplot(2,2,2)
plot(x,y2, ':')
subplot(2,2,[3 4])
plot(x,y3, 'o')
```

Other specialized planar plotting functions you may wish to explore via help are:

bar	feather	hist	quiver	
compass	fill	polar	rose	stairs

15.10 Graphics hard copy

Select File ▶Print or click the print button

in the Figure window to send a copy of your figure to your default printer. Layout options and selecting a printer can be done with File ▶Page Setup and File ▶Print Setup.

You can save the figure as a file for later use in a MATLAB Figure window. Try the save button

or File ▶Save. This saves the figure as a .fig file, which can be later opened in the Figure window with the open button

or with File ▶Open. Selecting File ▶Export Setup or File ▶Save As allows you to convert your figure to many other formats.

16 Three-Dimensional Graphics

Primary commands for creating three-dimensional graphics of numerically-defined functions are plot3, mesh, surf, and light. Plotting of symbolic functions is discussed in Chapter 19. The menu options and commands for setting axes, scaling, and placing text, labels, and legends on a graph also apply for 3-D graphs. A zlabel can be added. The axis command requires a vector of length 6 with a 3-D graph.

16.1 Curve plots

Completely analogous to plot in two dimensions, the command plot3 produces curves in three-dimensional space. If x, y, and z are three vectors of the same size, then the command plot3(x,y,z) produces a perspective plot of the piecewise linear curve in three-space passing through the points whose coordinates are the respective elements of x, y, and z. These vectors are usually defined parametrically. For example,

```
t = .01:.01:20*pi ;
x = cos(t) ;
y = sin(t) ;
z = t.^3 ;
plot3(x, y, z)
```

produces a helix that is compressed near the *x-y* plane (a "slinky"). Try it.

16.2 Mesh and surface plots

The mesh command draws three-dimensional wire mesh surface plots. The command mesh(z) creates a three-dimensional perspective plot of the elements of the matrix z. The mesh surface is defined by the *z*-coordinates of points above a rectangular grid in the *x-y* plane. Try mesh(eye(20)).

Similarly, three-dimensional faceted surface plots are drawn with the command surf. Try surf(eye(20)).

To draw the graph of a function $z = f(x, y)$ over a rectangle, first define vectors xx and yy, which give partitions of the sides of the rectangle. Then [x,y]=meshgrid(xx,yy) creates a matrix x, each row of which equals xx (whose column length is the length of yy) and similarly a matrix y, each column of which equals yy. A matrix z, to which mesh or surf can be applied, is then computed by evaluating the function f entry-wise over the matrices x and y. You can, for example, draw the graph of $z = e^{-x^2-y^2}$ over the square $[-2,2] \times [-2,2]$ as follows:

```
xx = -2:.2:2 ;
yy = xx ;
[x, y] = meshgrid(xx, yy) ;
z = exp(-x.^2 - y.^2) ;
mesh(z)
```

Try this plot with `surf` instead of `mesh`. Note that you must use `x.^2` and `y.^2` instead of `x^2` and `y^2` to ensure that the function acts entry-wise on `x` and `y`.

16.3 Parametrically defined surfaces

Plots of parametrically defined surfaces can also be made. See the MATLAB functions `sphere` and `cylinder` for example. The next example displays the cover of this book, with lighting, color, and viewpoint defined in Section 16.6. First, start a figure and set up the mesh:

```
figure(1) ; clf
t = linspace(0, 2*pi, 512) ;
[u,v] = meshgrid(t) ;
```

Next, define the surface:[1]

```
a = -0.2 ; b = .5 ; c = .1 ;
n = 2 ;
x = (a*(1-v/(2*pi)).*(1+cos(u)) + c) ...
    .* cos(n*v) ;
y = (a*(1-v/(2*pi)).*(1+cos(u)) + c) ...
    .* sin(n*v) ;
z = b*v/(2*pi) + ...
    a*(1-v/(2*pi)) .* sin(u) ;
```

Plot the surface, using `y-2*x` to define the color, and turn off the mesh lines on the surface:

```
surf(x,y,z,y-2*x)
shading interp
```

Also try `a=-0.5`, which gives the back cover.

Other three-dimensional plotting functions you may wish to explore via `help` or `doc` are `meshz`, `surfc`, `surfl`,

---

[1] von Seggern, CRC Standard Curves and Surfaces, 2nd ed., CRC Press, 1993, pp. 306-307.

contour, and pcolor. For plotting symbolically defined parametric surfaces (including the same seashell you plotted above), see Section 19.7.

16.4   Volume and vector visualization

MATLAB has an extensive suite of volume and vector visualization tools. The following example evaluates a function of three variables, v=f(x,y,z), that represents a fluid flow problem. It returns both v and the coordinates (x, y, and z) at which the function was evaluated.

```
[x,y,z,v] = flow ;
```

Now try visualizing it. The first method plots the surface at which v is -3; the second plots slices of the data:

```
figure(1) ; clf
isosurface(x, y, z, v, -3)
figure(2) ; clf
slice(x, y, z, v, [3 8], 0, 0)
```

Type doc specgraph for more volume and vector visualization tools.

16.5   Color shading and color profile

The color shading of surfaces is set by the shading command. There are three settings for shading: faceted (default), interpolated, and flat. These are set by the commands:

```
shading faceted
shading interp
shading flat
```

Note that on surfaces produced by surf, the settings interpolated and flat remove the superimposed mesh lines. Experiment with various shadings on the surface produced above. The command shading (as well as colormap and view described below) should be entered after the surf command.

The color profile of a surface is controlled by the colormap command. Available predefined color maps include hsv (the default), hot, cool, jet, pink, copper, flag, gray, bone, prism, and white. For example, colormap(cool) sets the cool color profile for the current figure. Experiment with various color maps on the surface produced above. See also doc colorbar.

16.6   Perspective of view

The Figure window provides a wide range of controls for viewing the figure. Select View ►Camera Toolbar to see these controls, or pull down the Tools menu. Try, for example, selecting Tools ►Rotate 3D, and then click the mouse in the Figure window and drag it to rotate the object. Some of these options can be controlled by the view and rotate3d commands, respectively.

The MATLAB function peaks generates an interesting surface on which to experiment with shading, colormap, and view. Type peaks, select Tools ►Rotate 3D, and click and drag the figure to rotate it.

In MATLAB, light sources and camera position can be set. Taking the peaks surface from the example above, select Insert ►Light, or type light to add a light source. See the online document *MATLAB 7 Graphics* for camera and lighting help.

This example defines the color, shading, lighting, surface material, and viewpoint for the cover of the book:

```
axis off
```

```
axis equal
colormap(hsv(1024))
shading interp
material shiny
lighting gouraud
lightangle(80, -40)
lightangle(-90, 60)
view([-120 0])
```

17 Advanced Graphics

MATLAB possesses a number of other advanced graphics
capabilities. Significant ones are bitmapped images,
object-based graphics (called Handle Graphics®), and
Graphical User Interface (GUI) tools.

17.1 Handle Graphics

Beyond those just described, the MATLAB graphics system
provides low-level functions that let you control virtually all
aspects of the graphics environment to produce
sophisticated plots. The commands set and get allow
access to all the properties of your plots. Try set(gcf) to
see some of the properties of a figure that you can control.
set(gca) lists the properties of the current axes (see
Section 17.3 for an example). This system is called Handle
Graphics®. See *MATLAB 7 Graphics* for more information.

17.2 Graphical user interface

The graphics system in MATLAB also provides the ability
to add sliders, push-buttons, menus, and other user interface
controls to your own figures. For information on creating
user interface controls, try doc uicontrol. This allows
you to create interactive graphical-based applications. GUIs

use *callback* functions, which are called whenever a user-interface object is triggered (when the user presses a button, for example).

Try guide (short for Graphic User Interface Development Environment). This brings up the Layout Editor window that you can use to interactively design a graphic user interface. Also see the online document *Creating Graphical User Interfaces*.

17.3 Images

The image function plots a matrix, where each entry in the matrix defines the color of a single pixel or block of pixels in the figure. image(K) paints the (i,j)th block of the figure with color K(i,j) taken from the colormap. Here is an example of the Mandelbrot set. The bottom left corner is defined as (x0,y0), and the upper right corner is (x0+d,y0+d). Try changing x0, y0, and d to explore other regions of the set (x0=-.38, y0=.64, d=.01 is also very pretty). This is also a good example of one-dimensional indexing:

```
x0 = -2 ; y0 = -1.5 ; d = 3 ; n = 512 ;
maxit = 256 ;
x = linspace(x0, x0+d, n) ;
y = linspace(y0, y0+d, n) ;
[x,y] = meshgrid(x, y) ;
C = x + y*1i ;
Z = C ;
K = ones(n, n) ;
for k = 1:maxit
    a = find((real(Z).^2+imag(Z).^2) < 4);
    Z(a) = (Z(a)).^2 + C(a) ;
    K(a) = k ;
end
figure(1) ; clf
```

119

```
colormap(jet(maxit)) ;
image(x0 + [0 d], y0 + [0 d], K) ;
set(gca, 'YDir', 'normal') ;
axis equal
axis tight
```

image, by default, reverses the y direction and plots the
K(1,1) entry at the top left of the figure (just like the spy
function described in Section 18.5). The set function resets
this to the normal direction, so that K(1,1) is plotted in the
bottom left corner. Try replacing the fourth argument in
surf, for the seashell example, with K, to paint the seashell
surface with the Mandelbrot set.

18   Sparse Matrix Computations

A sparse matrix is one with mostly zero entries. MATLAB
provides the capability to take advantage of the sparsity of
matrices.

18.1   Storage modes

MATLAB has two storage modes, full and sparse, with full
the default. Currently, only double or logical vectors or
two-dimensional arrays can be stored in the sparse mode.
The functions full and sparse convert between the two
modes. Nearly all MATLAB operators and functions
operate seamlessly on both full and sparse matrices. For a
matrix A, full or sparse, nnz(A) returns the number of
nonzero elements in A.

An m-by-n sparse matrix is stored as a set of sparse
columns, where each column is represented as a packed list
of nonzero values and their row indices. Thus, if A is sparse,
x=A(9,:) takes much more time than x=A(:,9), and
s=A(4,5) is also slow. To get high performance when

dealing with sparse matrices, use matrix expressions instead
of `for` loops and vector or scalar expressions. If you must
operate on the rows of a sparse matrix A, compute the
transpose (C=A') and work with the columns of C instead.

If a full tridiagonal matrix F is created via, say,

```
F = randi([0 9], 6)
F = triu(tril(F,1), -1)
```

then the statement S=sparse(F) converts F to sparse
mode. Try it. Note that the output lists the nonzero entries
in column major order along with their row and column
indices because of how sparse matrices are stored. The
statement F=full(S) returns F in full storage mode. You
can check the storage mode of a matrix A with the command
issparse(A).

18.2 Generating sparse matrices

A sparse matrix is usually generated directly rather than by
applying the function `sparse` to a full matrix. A sparse
banded matrix can be easily created via the function
`spdiags` by specifying diagonals. For example, a familiar
sparse tridiagonal matrix is created by:

```
m = 6 ;
n = 6 ;
e = ones(n,1) ;
d = -2*e ;
T = spdiags([e d e], [-1 0 1], m, n)
```

Try it. The integral vector [-1 0 1] specifies in which
diagonals the columns of [e d e] should be placed (use
full(T) to see the full matrix T and spy(T) to view T
graphically). Experiment with other values of m and n and,
say, [-3 0 2] instead of [-1 0 1]. See doc spdiags
for further features of spdiags.

The sparse analogs of `eye`, `zeros`, and `rand` for full matrices are, respectively, `speye`, `sparse`, and `sprand`. The `spones` and `sprand` functions take a matrix argument and replace only the nonzero entries with ones and uniformly distributed random numbers, respectively. `sparse(m,n)` creates a sparse zero matrix. `sprand` also permits the sparsity structure to be randomized. This is a useful method for generating simple sparse test matrices, but be careful. Random sparse matrices are not truly "sparse" because they experience catastrophic fill-in when factorized. Sparse matrices arising in real applications typically do not share this characteristic (see www.cise.ufl.edu/research/sparse/matrices).

The versatile function `sparse` also permits creation of a sparse matrix via listing its nonzero entries:

```
i = [1 2 3 4 4 4] ;
j = [1 2 3 1 2 3] ;
s = [5 6 7 8 9 10] ;
S = sparse(i, j, s, 4, 3)
full(S)
```

The last two arguments to `sparse` in the example above are optional. They tell `sparse` the dimensions of the matrix; if not present, then S is `max(i)`-by-`max(j)`. If there are repeated entries in `[i j]`, then the entries are added together. The commands below create a matrix whose diagonal entries are 2, 1, and 1.

```
i = [1 2 3 1] ;
j = [1 2 3 1] ;
s = [1 1 1 1] ;
S = sparse(i, j, s)
full(S)
```

The entries in `i`, `j`, and `s` can be in any order (the same order for all three arrays, of course).

In general, if the vector s lists the nonzero entries of S and the integral vectors i and j list their corresponding row and column indices, then sparse(i,j,s,m,n) creates the desired sparse m-by-n matrix S. As another example try:

```
n = 6 ;
e = randi([0 9], n-1, 1) ;
E = sparse(2:n, 1:n-1, e, n, n)
```

Creating a sparse matrix by assigning values to it one at a time is exceedingly slow; never do it if you can avoid it. The next example constructs the same matrix as A=sparse(i,j,s,m,n) (except for handling duplicate entries), but it should never be used because it is so slow:

```
A = sparse(m,n) ;
for k = 1:length(s)
    A(i(k),j(k)) = s(k) ;
end
```

18.3 Computation with sparse matrices

The arithmetic operations and most MATLAB functions can be applied independent of storage mode. The storage mode of the result depends on the storage mode of the operands or input arguments. Operations on full matrices always give full results. If F is a full matrix (not a scalar), S and Z are sparse matrices, and n is a (full) scalar, then these operations give sparse results:

S+S	S*S	S.*S	S.*F
S-S	S/Z	S\Z	-S
S'	S.'	inv(S)	chol(S)
lu(S)	diag(S)	max(S)	sum(S)
S*n	S/n	S^n	S.^n
n\S			

These give full results (even if n is full(0)):

```
S+F          F\S          S/F          S+n
S*F          S\F          F/S          S-n
```

A matrix built from blocks, such as [A, B; C, D], is
stored in sparse mode if any constituent block is sparse. To
compute the eigenvalues or singular values of a sparse
matrix S, you must convert S to a full matrix and then use
eig or svd, as eig(full(S)) or svd(full(S)). If S is a
large sparse matrix and you wish only to compute some of
the eigenvalues or singular values, then you can use the
eigs or svds functions (eigs(S) or svds(S)).

18.4   Permutation vectors and matrices

Permuting a matrix in MATLAB is often essential for
obtaining good performance and reliable numerical results.
A permutation can be represented as an index vector or as a
sparse matrix. Try this example, which uses the west0479
test matrix included in MATLAB.

```
load west0479 ; A = west0479 ;
[L,U,P,Q] = lu(A) ;
[L,U,p,q] = lu(A, 'vector') ;
```

Both commands return an LU factorization so that
$LU = PAQ$, where P is the row permutation and Q is the
column permutation. The permutation matrices are P and Q,
so that L*U equals P*A*Q. The permutation vectors are p
and q, so that L*U equals A(p,q). Sometimes you may
have one representation of a permutation (vector or matrix)
and need to compute the other one. Here are the rules for
conversion, where A is m-by-n.

```
[p j x] = find(P')      converts P*A to A(p,:)
[q j x] = find(Q)       converts A*Q to A(:,q)
P=sparse(1:m, p, 1)     converts A(p,:) to P*A
Q=sparse(q, 1:n, 1)     converts A(:,q) to A*Q
```

124

18.5 Visualizing matrices

The spy function introduced in the last section plots the
nonzero pattern of a sparse matrix. spy can also be used on
full matrices. It is useful for matrix expressions coming
from relational operators. Try this example (see Chapter 6
for the ddom function):

```
A = [
-1  2  3 -4
 0  2 -1  0
 1  2  9  1
-3  4  1  1]
C = ddom(A)
figure(1) ; spy(A ~= C)
figure(2) ; spy(A > 2)
```

What you see is a picture of where A and C differ, and
another picture of which entries of A are greater than 2.

19 The Symbolic Math Toolbox

The Symbolic Math Toolbox extends the numeric and
graphic power of MATLAB by adding the capability of
computing and manipulating symbolic mathematical
expressions. The Symbolic Math Toolbox is included in the
Student Version of MATLAB. Since the Symbolic Math
Toolbox is not part of the Professional Version of MATLAB
(by default), it may not be installed on your system, in
which case this chapter will not apply.

Many of the functions in the Symbolic Math Toolbox have
the same names as their numeric counterparts. MATLAB
selects the correct one depending on the type of inputs to
the function. Typing doc eig and doc symbolic/eig
displays the help for the numeric eigenvalue function and its
symbolic counterpart, respectively.

19.1 Symbolic variables

You can declare a variable as symbolic with the `syms` statement. For example,

```
syms x
```

creates a symbolic variable x. The statement:

```
syms x real
```

declares to MATLAB that x is a symbolic variable with no imaginary part.

You can also assert to MATLAB that x is always positive, with `syms x positive`. To clear the `real` status of x, use `syms x clear`. Now MATLAB must assume that x can be either real or complex.

Symbolic computations are handled by a separate symbolic engine called *MuPAD*, which keeps track of all your symbolic variables. The statement `clear x` deletes the variable from the MATLAB workspace, but not the MuPAD workspace. If you declare x as `real`, and then `clear` it and recreate it, MuPAD thinks that x is still limited to `real`. Thus, use `syms x clear` to clear the `real` status of x, or use `reset(symengine)` to reset the MuPAD symbolic engine and return all symbolic variables to their defaults. The statement `clear all` clears the MATLAB workspace and also resets the MuPAD symbolic engine.

Symbolic variables can be constructed from existing numeric variables using the `sym` function. Try:

```
z = 1/10
a = sym(z)
y = rand(1)
b = sym(y, 'd')
```

although better ways to create a include:

```
a = sym('1/10')
a = 1 / sym(10)
```

If you want to ensure a precise symbolic expression, you must avoid numeric computations. Compare these three expressions. The first is only accurate to double-precision numeric computation (about 16 digits). The second and third avoid numeric computation completely.

```
sym(log(2))
sym('log(2)')
log(sym(2))
```

You can create a symbolic abstract function. This example declares f(x) as some unknown function of x:

```
syms x
f = sym('f(x)')
```

The syms command and sym function have many more options. See doc syms and doc sym.

19.2 Calculus

The function diff computes the symbolic derivative of a function defined by a symbolic expression. First, to define a symbolic expression, you should create symbolic variables and then proceed to build an expression as you would mathematically. For example,

```
syms x
f = x^2 * exp(x)
diff(f)
```

creates a symbolic variable x, builds the symbolic expression $f = x^2 e^x$, and returns the symbolic derivative of f with respect to x: x^2*exp(x) + 2*x*exp(x) in MATLAB notation. Try it. Next,

```
syms t
diff(sin(pi*t))
```

returns the derivative of $\sin(\pi t)$, as a function of t.

Here are examples of taking the derivative of an abstract function, illustrating the product, quotient, and reciprocal rules of calculus, and a special case of the chain rule. The function pretty displays a symbolic expression in an easier-to-read form resembling typeset mathematics. See Section 19.5 for simple.

```
syms x n
f = sym('f(x)')
g = sym('g(x)')
pretty(diff(f*g))
pretty(diff(f/g))
pretty(diff(1/f))
pretty(simple(diff(f^n)))
```

Formats in addition to pretty include latex, ccode, and fortran. Try, for example,

```
syms x a b
f = x/(a*x+b)
pretty(f)
g = int(f)
pretty(g)
latex(g)
ccode(g)
fortran(g)
int(g)
pretty(ans)
```

Partial derivatives can also be computed. Try:

```
syms x y
g = x*y + x^2
diff(g)                 computes δg/δx
```

```
diff(g, x)              also computes δg/δx
diff(g, y)              computes δg/δy
```

To permit omission of the second argument for functions
such as the above, MATLAB chooses a default symbolic
variable for the symbolic expression. The findsym
function returns the default choice. Its rule is to choose the
variable whose name is nearest x in the alphabet. You can,
of course, override the default choice as shown above. Try,
for example,

```
syms x x1 x2 theta
F = x * (x1*x2 + x1 - 2)
findsym(F,1)
diff(F, x)              computes δF/δx
diff(F, x1)             computes δF/δx₁
diff(F, x2)             computes δF/δx₂
G = cos(theta*x)
diff(G, theta)          computes δG/δθ
```

diff can compute second or higher-order derivatives. The
second derivative of $\sin(2x)$ is given by either of the
following two examples:

```
diff(sin(2*x), 2)
diff(sin(2*x), x, 2)
```

With a numeric argument, diff is the difference operator
of basic MATLAB, which can be used to numerically
approximate the derivative of a function. See doc diff or
help diff for the numeric function, and doc
symbolic/diff or help sym/diff for the symbolic
derivative function.

The function int attempts to compute the indefinite integral
(antiderivative) of a function defined by a symbolic
expression. Try, for example,

```
syms a b t x y z theta
int(sin(a*t + b))
int(sin(a*theta + b), theta)
int(x*y^2 + y*z, y)
int(x^2 * sin(x))
```

Note that, as with `diff`, when the second argument of `int`
is omitted, the default symbolic variable (as selected by
`findsym`) is chosen as the variable of integration. In some
instances, `int` is unable to give a result in terms of
elementary functions. Consider, for example,

```
int(exp(-x^2))
int(sqrt(1 + x^3))
```

In the first case the result is given in terms of the error
function `erf`, whereas in the second, the result is given in
terms of `ellipticF`, a function defined by an integral.

Here is a basic integral rule with an abstract function:

```
f = sym('f(x)')
int(diff(f) / f)
```

Definite integrals can also be computed by using additional
input arguments. Try, for example,

```
int(sin(x), 0, pi)
int(sin(theta), theta, 0, pi)
```

In the first case, the default symbolic variable x was used as
the variable of integration to compute:

$$\int_0^\pi \sin x\, dx$$

whereas, in the second, `theta` was chosen. Other definite
integrals you can try are:

130

```
int(x^5, 1, 2)
int(log(x), 1, 4)
int(x * exp(x), 0, 2)
int(exp(-x^2), 0, inf)
```

It is important to realize that the results returned are symbolic expressions, not numeric ones. The function double converts these into MATLAB floating-point numbers, if desired. For example, the result returned by the first integral above is 21/2. Entering double(ans) then returns the MATLAB numeric result 10.5000.

Alternatively, you can use the function vpa (variable precision arithmetic; see Section 19.3) to convert the expression into a symbolic number of arbitrary precision. For example,

```
int(exp(-x^2), 0, inf)
```

gives the result:

```
pi^(1/2)/2
```

Then the statement:

```
vpa(ans, 25)
```

symbolically gives the result to 25 significant digits:

```
.8862269254527580136490837
```

You may wish to contrast these techniques with the MATLAB numerical integration functions quad and quadl (see Section 20.4).

The limit function is used to compute the symbolic limits of various expressions. For example,

```
syms h n x
limit((1 + x/n)^n, n, inf)
```

computes $\lim_{n \to \infty} (1 + x/n)^n$ as (`inf` represents ∞ in a MATLAB). You should also try:

```
limit(sin(x), x, 0)
limit((sin(x+h)-sin(x))/h, h, 0)
```

The `taylor` function computes the Maclaurin and Taylor series of symbolic expressions. For example,

```
taylor(cos(x) + sin(x))
```

returns the fifth order Maclaurin polynomial approximating $\cos x + \sin x$. This returns the eighth degree Taylor approximation to $\cos x^2$ centered at the point $x_0 = \pi$:

```
taylor(cos(x^2), 8, x, pi)
```

19.3   Variable precision arithmetic

Three kinds of arithmetic operations are available:

numeric	floating-point arithmetic in MATLAB
rational	exact symbolic arithmetic in MuPAD
VPA	variable precision arithmetic in MuPAD

One can obtain exact rational results with, for example,

```
s = simple(sym('13/17 + 17/23'))
```

You are already familiar with numeric computations. For example, with `format long`,

```
pi*log(2)
```

gives the numeric result `2.17758609030360`.

Numeric computations in MATLAB are done in approximately 16 decimal digit floating-point arithmetic. With `vpa`, you can obtain results to arbitrary precision, within the limitations of time and memory. Try:

```
vpa('pi * log(2)')
vpa(sym(pi) * log(sym(2)))
vpa('pi * log(2)', 50)
```

The default precision for vpa is 32. Hence, the two results
are accurate to 32 digits, whereas the third is accurate to the
specified 50 digits. Ludolf van Ceulen (1540-1610)
calculated π to 36 digits. The Symbolic Math Toolbox can
quite easily compute π to 10,000 digits or more. Try:

```
pretty(vpa('pi', 10000))
```

The default precision can be changed with the function
digits. While the rational and VPA computations can be
more accurate, they are in general slower than numeric
computations. If you pass a numeric expression to vpa,
MATLAB evaluates it numerically first, so use a symbolic
expression or place the expression in quotes. Compare your
results, above, with:

```
vpa(pi * log(2))
```

which is accurate to only about 16 digits (even though 32
digits are displayed). This is a common mistake with the
use of vpa and the Symbolic Math Toolbox in general.

19.4 Numeric and symbolic substitution

Once you have a symbolic expression, you can modify it or
evaluate it numerically with the subs function. The
function subs replaces all occurrences of the symbolic
variable in an expression by a specified second expression.
This corresponds to the composition of two functions. Try,
for example,

```
syms x s t
subs(sin(x), x, pi/3)
subs(sin(x), x, sym(pi)/3)
```

```
double(ans)
subs(g*t^2/2, t, sqrt(2*s))
subs(sqrt(1-x^2), x, cos(x))
subs(sqrt(1-x^2), 1-x^2, cos(x))
```

The general idea is that in the statement
subs(expr,old,new) the third argument (new) replaces
the second argument (old) in the first argument (expr).
Compare the first two examples above. The result is
numeric if all variables in the expression are substituted
with numeric values, or symbolic otherwise.

You can substitute multiple symbolic expressions, numeric
expressions, or any combination, using cell arrays of
symbolic or numeric values. Try:

```
syms x y
S = x^y
subs(S, x, 3)
subs(S, {x y}, {3 2})
subs(S, {x y}, {3 x+1})
```

You can perform multiple substitutions for any one
symbolic variable, which returns a matrix of symbolic
expressions or numeric values. Try this, which constructs a
function F, finds its derivative G, and evaluates G at
x=0:.1:1.

```
syms x
F = x^2 * sin(x)
G = diff(F)
subs(G, x, 0:.1:1)
```

Also try:

```
a = subs(S, y, 1:9)
a(3)
a = subs(S, {x y},{2*ones(9,1) (1:9)'})
```

The first expression returns a row vector containing the symbolic expressions x, x^2, ... x^9. The second substitution returns a numeric column vector containing the powers of 2 from 2 to 512. Each entry in the cell array must be of the same size.

Substitution acts just like composition in calculus. Taking the derivative of function composition illustrates the chain rule of calculus:

```
f = sym('f(x)')
g = sym('g(x)')
diff(subs(f, g))
pretty(ans)
```

19.5   Algebraic simplification

Convenient algebraic manipulations of symbolic expressions are available.

The function expand distributes products over sums and applies other identities, whereas factor attempts to do the reverse. The function collect views a symbolic expression as a polynomial in its symbolic variable (which may be specified) and collects all terms with the same power of the variable. To explore these capabilities, try the following:

```
syms a b x y z
expand((a + b)^5)
factor(ans)
expand(exp(x + y))
expand(sin(x + 2*y))
factor(x^6 - 1)
collect(x * (x * (x + 3) + 5) + 1)
horner(ans)
collect((x + y + z)*(x - y - z))
```

```
collect((x + y + z)*(x - y - z), y)
collect((x + y + z)*(x - y - z), z)
diff(x^3 * exp(x))
factor(ans)
```

The powerful function `simplify` applies many identities in
an attempt to reduce a symbolic expression to a simple
form. Try, for example,

```
simplify(sin(x)^2 + cos(x)^2)
simplify(exp(5*log(x) + 1))
d = diff((x^2 + 1)/(x^2 - 1))
simplify(d)
```

The alternate function `simple` computes several
simplifications and chooses the shortest of them. It often
gives better results on expressions involving trigonometric
functions. Try the following commands:

```
simplify(cos(x) + (-sin(x)^2)^(1/2))
simple   (cos(x) + (-sin(x)^2)^(1/2))
simplify((1/x^3+6/x^2+12/x+8)^(1/3))
simple   ((1/x^3+6/x^2+12/x+8)^(1/3))
```

The function `factor` can also be applied to a numeric or
symbolic integer argument to compute the prime
factorization of the integer. Try, for example,

```
factor(4248)
factor(sym('4248'))
factor(sym('4549319348693'))
factor(sym('4549319348597'))
```

19.6   Two-dimensional graphs

The MATLAB function `fplot` (see Section 15.3) provides
a tool to conveniently plot the graph of a function. Since it
is, however, the name or handle of the function to be plotted

136
```
```

that is passed to `fplot`, the function must first be defined in an M-file (or else be a built-in function or anonymous function).

In the Symbolic Math Toolbox, `ezplot` lets you plot the graph of a function directly from its defining symbolic expression. For example, to plot a function of one variable try:

```
syms t x y
ezplot(sin(2*x))
ezplot(t + 3*sin(t))
ezplot(2*x/(x^2 - 1))
ezplot(1/(1 + 30*exp(-x)))
```

By default, the x-domain is `[-2*pi, 2*pi]`. This can be overridden by a second input variable, as with:

```
ezplot(x*sin(1/x), [-.2 .2])
```

You will often need to specify the x-domain and y-domain to zoom in on the relevant portion of the graph. Compare, for example,

```
ezplot(x*exp(-x))
ezplot(x*exp(-x), [-1 4])
```

`ezplot` attempts to make a reasonable choice for the y-axis. With the last figure, select Edit ▶Axes Properties in the Figure window and modify the y-axis to start at `-3`, and hit enter. Changing the x-axis in the Property Editor does not cause the function to be reevaluated, however.

To plot an implicitly defined function of two variables, try this:

```
ezplot(x^2 + y^2 - 1)
```

which plots the unit circle over the default x-domain and y-domain of `[-2*pi, 2*pi]`. Since this is too large for the unit circle, try this instead:

```
ezplot(x^2 + y^2 - 1, [-1 1 -1 1])
```

The first two entries in the second argument define the
x-domain. The second two define the *y*-domain. If the
y-domain is the same as the *x*-domain, then you only need
to specify the *x*-domain (see the next example).

In both of the previous examples, you plotted a circle but it
looks like an ellipse. This is because with auto-scaling, the
x and *y* axes are not equal. Fix this by typing:

```
axis equal
```

To plot a parametrized function, provide two function
arguments. Try this, which plots a cycloid over the domain
-4π to 4π.

```
x = t-sin(t)
y = 1-cos(t)
ezplot(x,y, [-4*pi 4*pi])
```

The `ezpolar` function creates polar plots. Try creating a
three-leaf rose and a hyperbolic spiral:

```
ezpolar(sin(3*t))
ezpolar(1/t, [1 10*pi])
```

Entering the command `funtool` (no input arguments)
brings up three graphic figures, two of which will display
graphs of functions and one containing a control panel. This
function calculator lets you manipulate functions and their
graphs for pedagogical demonstrations. Type `doc funtool`
for details.

19.7 Three-dimensional surface graphs

MATLAB has several easy-to-use functions for creating
three-dimensional surface graphs.

ezcontour	3-D contour plot
ezcontourf	3-D filled contour plot
ezmesh	3-D mesh plot
ezmeshc	3-D mesh and contour plot
ezsurf	3-D surface plot
ezsurfc	3-D surface and contour plot

Here is an interesting function to try:

```
syms x y
f = sin((x^2+y)/2)/(x^2-x+2)
ezsurfc(f)
```

Try each of these plotting functions with this function f. For this function, ezcontourf gives more information than ezcontour because the function fluctuates across a single contour in several regions. The default domain for x and y is -2π to 2π. You can change this with an optional second parameter. Try:

```
ezsurf(f, [-4 4 -pi pi])
```

which defines the x-domain as -4 to 4, and the y-domain as $-\pi$ to π The appearance of the plots can be modified by the shading command after the figure is plotted (see Section 16.5).

Functions with discontinuities or singularities can cause difficulty for these graphing functions. Here is an example that is similar to the function f above,

```
f = sin(abs(sqrt(x^2+y)))/(x^2-x+2)
ezsurf(f)
```

Click the rotate button

in the Figure window, then click and drag the graph itself. The function touches the $z = 0$ plane along the curve

139

defined by $y = -x^2$, but the graph does not capture this property very well because the gradient is not defined along that curve. To plot this function accurately, you would need to define your own mesh points, compute the function numerically, and use `surf` or another numerical graphing function instead.

The four mesh and surface functions listed above can also plot parametrized surface functions. The first three arguments are the $x(s,t)$, $y(s,t)$, and $z(s,t)$ functions, and the last (optional) argument defines the domain. To create a symbolic seashell, start a new figure and define your symbolic variables:

```
figure(1) ; clf
syms u v x y z
```

Next, define x, y, and z, just as you did for the numeric seashell in Section 16.3 (do not use the `linspace` and `meshgrid` commands). The MATLAB statements are the same, except that now these variables are defined symbolically, not numerically. Plot the symbolic surface:

```
ezsurfc(x,y,z,[0 2*pi])
```

Turn off the axis and set the shading, material, lighting, color, and viewpoint, just as in Sections 16.3 and 16.6.

19.8   Three-dimensional curves

Parameterized 3-D curves are plotted with `ezplot3`. Try this example, which combines a folium of Descartes in the x-y plane with a sinusoid in the z direction:

```
syms x y z t
x = 3*t / (1+t^3)
y = 3*t^2 / (1+t^3)
z = sin(t)
ezplot3(x,y,z)
```

The default domain for *t* is 0 to 2π. Here is an example of how to change it:

```
ezplot3(x,y,z,[-.9 10])
```

The ezplot3 function can animate the plot so that you can observe how *x*, *y*, and *z* depend on *t*. Try both of these examples. The ball moves quickly over the first half of the curve but more slowly over the second half:

```
ezplot3(x,y,z,'animate')
ezplot3(x,y,z, [-.9 10], 'animate')
```

The 2-D curve plotting function ezplot cannot animate its plot, but you can do the same with ezplot3. Just give it a *z* argument of zero. Try:

```
syms z
z = 0
ezplot3(x,y,z,'animate')
```

and then rotate the graph so that you are viewing the *x-y* plane. Click the rotate button and drag the graph, or right-click the graph and select Go to X-Y view. Then click the Repeat button in the bottom left corner.

19.9   Symbolic matrix operations

This toolbox lets you represent matrices in symbolic form as well as numeric form. Given numeric matrix a, sym(a) converts a to a symbolic matrix. Try:

```
a = magic(3)
A = sym(a)
```

The function double(A) converts the symbolic matrix back to a numeric one.

Symbolic matrices can also be generated directly. Try, for example,

```
syms a b s
K = [a + b, a - b ; b - a, a + b]
G = [cos(s), sin(s); -sin(s), cos(s)]
```

Here G is a symbolic Givens rotation matrix.

Algebraic matrix operations with symbolic matrices are computed as you would in MATLAB:

K+G	matrix addition
K-G	matrix subtraction
K*G	matrix multiplication
K\G	left matrix division
K/G	right matrix division
G^2	power
G.'	array transpose
G'	matrix transpose

These operations are illustrated by the following, which use the matrices K and G generated above. The last expression demonstrates that G is orthogonal.

```
L = K^2
collect(L)
factor(L)
diff(L, a)
int(K, a)
J = K/G
simplify(J*G)
simplify(G*(G.'))
```

The initial result of the basic operations may not be in the form desired for your application; so it may require further processing with simplify, collect, factor, or expand. These functions, as well as diff and int, act entry-wise on a symbolic matrix.

19.10 Symbolic linear algebraic functions

In addition to the MATLAB matrix operators, some of the functions that are useful for symbolic computation include:

det	determinant
inv	inverse
null	basis for null space
colspace	basis for column space
eig	eigenvalues and eigenvectors
poly	characteristic polynomial
svd	singular value decomposition
jordan	Jordan canonical form

These functions take either symbolic or numeric matrices as inputs. Computations with symbolic rational matrices are carried out exactly. The inv and det functions should rarely be used for numeric matrices, but work well for symbolic matrices. Try, for example,

```
c = randi([0 9], 4)
D = sym(c)
A = inv(D)
inv(A)
inv(A) * A
det(A)
b = ones(4,1)
x = A\b
A*x
A^3
```

For the matrices K and G defined in the previous section, try:

```
inv(K)
simplify(inv(G))
p = poly(G)
simplify(p)
pretty(simple(solve(p)))
```

```
pretty(simple(eig(G)))

y = simple(svd(G))
pretty(y)

syms s real
r = simple(svd(G))
syms s clear
```

Compare y and r. If you do not declare s as real, the svd of the 2-by-2 Givens rotation matrix does not demonstrate that the singular values are all equal to one.

A typical exercise in a linear algebra course is to determine those values of t so that, say,

```
A = [t 1 0 ; 1 t 1 ; 0 1 t]
```

is singular. The following simple computation:

```
syms t
A = [t 1 0 ; 1 t 1 ; 0 1 t]
p = det(A)
solve(p)
```

shows that this occurs for $t = 0$, $\sqrt{2}$, and $-\sqrt{2}$. See the next section (19.11) for the solve function.

The function eig attempts to compute the eigenvalues and eigenvectors in an exact closed form. Try, for example,

```
for n = 4:6
    A = sym(magic(n))
    [V, D] = eig(A)
end
```

Except in special cases, however, the result is usually too complicated to be useful. Try, for example, executing:

```
A = sym(randi([0 9], 3))
[V, D] = eig(A)
pretty(V)
```

a few times. The eigenvectors V are not very pretty. For this reason, it is usually more efficient to do the computation in variable-precision arithmetic, as is illustrated by:

```
A = vpa(randi([0 9], 3))
[V, D] = eig(A)
```

The comments above regarding eig apply as well to the computation of the singular values of a matrix by svd, as can be observed by repeating some of the computations above using svd instead of eig.

19.11 Solving algebraic equations

For a symbolic expression S, the statement solve(S) will attempt to find the values of the symbolic variable for which the symbolic expression is zero. The solve function cannot solve all equations. It does well with low-degree polynomial equations, but can have difficulty with trigonometric or other transcendental equations. If an exact symbolic solution is found, you can convert it to a floating-point solution via double. If an exact symbolic solution cannot be found, then a variable precision one is computed. Here are three similar equations. The first returns a symbolic result, the second a numeric vpa result, and the last one fails.

```
syms x b
solve(2^x - b)
solve(2^x + 3^x - 1)
solve(2^x + 3^x - b)
```

If you have an expression that contains several symbolic variables, you can solve for a particular variable by including it as an input argument in `solve`. The default variable solved for is x, or the one closest (alphabetically) to x if x is not a variable in the equation. Try this example:

```
syms x
f = x*cos(x) - 1
s = solve(f)
```

Here are some more examples:

```
Z = solve(x^2 + 2*x - 1)
pretty(Z)

syms x y z
f = x^2 + y^2 + z^2 + x*y*z
a = solve(f)
pretty(a)
simplify (subs (f, 'x', a))

b = solve(f, y)
pretty(b)
simplify (subs (f, 'y', b))
```

a is a solution in the variable x, and b is a solution in y. The inputs to `solve` can be quoted strings or symbolic expressions. To solve an equation whose right-hand side is not zero, use a quoted string or rearrange the equation:

```
X = solve('log(x) = x - 2')
X = solve(log(x) - x + 2)
vpa(X)
X = solve('2^x = x + 2')
X = solve(2^x - x - 2)
vpa(X)
```

This solves for the variable a:

```
solve('1 + (a+b)/(a-b) = b', 'a')
```

146

This solves the same for b, finding two solutions:

```
solve('1 + (a+b)/(a-b) = b', 'b')
```

The solution to the next example should be familiar. Try:

```
syms a b c x
solve(a*x^2 + b*x + c, x)
pretty(ans)
```

The function `solve` can also compute solutions of systems of general algebraic equations. To solve, for example, the nonlinear system below, it is convenient to first express the equations as strings.

```
S1 = 'x^2 + y^2 + z^2 = 2'
S2 = 'x + y = 1'
S3 = 'y + z = 1'
```

The solutions are then computed by:

```
[X, Y, Z] = solve(S1, S2, S3)
```

If you request the set of solutions in a single output with multiple unknowns, a `struct` is returned. Try

```
a = solve(S1, S2, S3)
a.x
a.y
a.z
```

If you alter S2 to:

```
S2 = 'x + y + z = 1'
```

then the solution computed by:

```
[X, Y, Z] = solve(S1, S2, S3)
```

will be given in terms of square roots. If you prefer solving symbolic expressions instead of strings, try

```
syms x y z
S1 = x^2 + y^2 + z^2 - 2
S2 = x + y - 1
S3 = y + z - 1
a = solve(S1, S2, S3)
```

The output of `solve` is in alphabetical order. For example, if you changed the name of z to w in these three equations the results would be returned in the order [W,X,Y]. The `solve` function can take quoted strings or symbolic expressions as input arguments, or a mixture of both.

19.12 Solving differential equations

The function `dsolve` solves ordinary differential equations. The symbolic differential operator is D:

```
Y = dsolve('Dy = x^2*y', 'x')
```

produces the solution `C2*exp(x^3/3)` to the differential equation $y' = x^2y$, where `C2` is some unspecified constant. The solution to an initial value problem can be computed by adding a second symbolic expression giving the initial condition.

```
Y = dsolve('Dy = x^2*y', 'y(0)=4', 'x')
```

Notice that in both examples above, the final input argument, `'x'`, is the independent variable of the differential equation. If no independent variable is supplied to `dsolve`, then it is assumed to be `t`. The higher order symbolic differential operators D2, D3, ... can be used to solve higher order equations. Try:

```
dsolve('D2y + y = 0')
dsolve('D2y + y = x^2', 'x')
dsolve('D2y + y = x^2', ...
    'y(0) = 4', 'Dy(0) = 1', 'x')
```

```
dsolve('D2y - Dy = 2*y')
dsolve('D2y + 6*Dy = 13*y')
dsolve('D3y - 3*Dy = 2*y')
pretty(ans)
```

Systems of differential equations can also be solved:

```
E1 = 'Dx = -2*x + y'
E2 = 'Dy = x - 2*y + z'
E3 = 'Dz = y - 2*z'
```

The solutions are then computed with:

```
[x, y, z] = dsolve(E1, E2, E3)
pretty(x)
pretty(y)
pretty(z)
```

You can explore further details with doc dsolve.

19.13 Further MuPAD access

In all of the examples you have worked through so far, you accessed the MuPAD symbolic engine through MATLAB commands. You can access more features in the Symbolic Math Toolbox via a powerful GUI interface called the MuPAD Notebook. Type mupadwelcome at the command line, and create a new Notebook. Once the GUI starts, you can enter commands directly into the MuPAD Notebook. The syntax of the MuPAD language is much like Pascal. It is very different than the MATLAB language and beyond the scope of this primer. For help, type doc(symengine) in the MATLAB Command Window (be sure to use the parentheses) or click the help button in the menu bar at the top of the MuPAD Notebook. For an extensive tutorial on MuPAD, click on the PDF documentation link in the MuPAD Help window, and read *The MuPAD Tutorial*.

20 Polynomials, Interpolation, and Integration

Polynomial functions are frequently used by numerical methods, and thus MATLAB provides several routines that operate on polynomials and piece-wise polynomials.

20.1 Representing polynomials

Polynomials are represented as vectors of their coefficients, so $f(x) = x^3 - 15x^2 - 24x + 360$ is simply

```
p = [1 -15 -24 360]
```

The roots of this polynomial (15, $\sqrt{24}$, and $-\sqrt{24}$):

```
r = roots(p)
```

Given a vector of roots r, poly(r) constructs the coefficients of the polynomial with those roots. With a little bit of roundoff error, you should see the coefficients of the original polynomial. Try it.

The poly function also computes the characteristic polynomial of a matrix whose roots are the eigenvalues of the matrix. The polynomial f(x) was chosen as the characteristic equation of the magic(3) matrix. Try:

```
A = magic(3)
s = poly(A)
roots(s)
eig(A)
f = poly(sym(A))
solve(f)
eig(sym(A))
```

20.2 Evaluating polynomials

You can evaluate a polynomial at one or more points with
the `polyval` function.

```
x = -1:2 ;
y = polyval(p,x)
```

Compare y with `x.^3-15*x.^2-24*x+360`. You can
construct a symbolic polynomial from the coefficient vector
p and back again:

```
syms x
f = poly2sym(p)
sym2poly(f)
```

20.3 Polynomial interpolation

Polynomials are useful as easier-to-compute
approximations of more complicated functions, via a Taylor
series expansion or by a low-degree best-fit polynomial
using the `polyfit` function. The statement:

```
p = polyfit(x, y, n)
```

finds the best-fit n-degree polynomial that approximates the
data points x and y. Try this example:

```
x = 0:.1:pi ;
y = sin(x) ;
p = polyfit(x, y, 5)
figure(1) ; clf
ezplot(@sin, [0 pi])
hold on
xx = 0:.001:pi ;
plot(xx, polyval(p,xx), 'r-')
```

Piecewise-polynomial interpolation is typically better than a
single high-degree polynomial. Try this example:

```
n = 10
x = -5:.1:5 ;
y = 1 ./ (x.^2+1) ;
p = polyfit(x, y, n)
figure(2) ; clf
ezplot(@(x) 1 ./ (x.^2+1))
hold on
xx = -5:.01:5 ;
plot(xx, polyval(p,xx), 'r-')
```

As n increases, the error in the center improves but increases dramatically near the endpoints. The spline and pchip functions compute piecewise-cubic polynomials which are better for this problem. Try:

```
figure(3) ; clf
yy = spline(x, y, xx) ;
plot(xx, yy, 'g')
```

Alternatively, with two inputs, spline and pchip return a struct that contains the piecewise polynomial, which can be later evaluated with ppval. Try:

```
figure(4) ; clf
pp = spline(x, y)
yy = ppval(pp, xx) ;
plot(xx, yy, 'c')
```

The spline function computes the conventional cubic spline, with a continuous second derivative. In contrast, pchip returns a piecewise polynomial with a discontinuous second derivative, but it preserves the shape of the function better than spline. pchip and spline have the same input and output arguments; they just compute different piecewise cubic polynomials that interpolate the data.

Polynomial multiplication and division (convolution and deconvolution) are performed by the conv and deconv functions. MATLAB also has a built-in fast Fourier transform function, fft.

20.4 Numeric integration (quadrature)

The quad and quadl functions are the numeric equivalent
of the symbolic int function, for computing a definite
integral. Both rely on polynomial approximations of
subintervals of the function being integrated. quadl is a
higher-order method that can be more accurate. The syntax
quad(@f,a,b) computes an approximate of the definite
integral,

$$\int_a^b f(x)\,dx$$

Compare these examples:

```
quad(@(x) x.^5, 1, 2)
quad(@log, 1, 4)
quad(@(x) x .* exp(x), 0, 2)
quad(@(x) exp(-x.^2), 0, 1e6)
quad(@(x) sqrt(1 + x.^3), -1, 2)
quad(@(x) real(airy(x)), -3, 3)
```

with the same results from the Symbolic Math Toolbox:

```
int('x^5', 1, 2)
int('log(x)', 1, 4)
int('x * exp(x)', 0, 2)
int('exp(-x^2)', 0, inf)
int('sqrt(1 + x^3)', -1, 2)
int('real(airy(x))', -3, 3)
```

Symbolic integration (int) can find a simple closed-form
solution to the first four examples, above. The fifth example
is not in closed form, and the last example cannot be solved
by int at all. It can only be computed numerically, with
quad.

The function f provided to quad and quadl must operate
on a vector x and return f(x) for each component of the

vector. An optional fourth argument to quad and quadl modifies the error tolerance. Double and triple integrals are evaluated by dblquad and triplequad. Array-valued functions are integrated with quadv.

21 Solving Equations

Solving equations is at the core of what MATLAB does. First, we will look back at the kinds of equations you have seen so far in the book. Next, in this chapter you will learn how MATLAB finds numerical solutions to nonlinear equations and systems of differential equations.

21.1 Symbolic equations

The Symbolic Math Toolbox can solve symbolic linear systems of equations using backslash (Section 19.9), nonlinear systems of equations using the solve function (Section 19.11), and systems of differential equations using dsolve (Section 19.12). The rest of MATLAB focuses on finding numeric solutions to equations, not symbolic.

21.2 Linear systems of equations

The pervasive and powerful backslash operator solves linear systems of equations of the form A*x=b (Sections 3.3, 18.3, and 19.9). The expression x=A\b handles the case when A is square or rectangular (under- or over-determined), full-rank or rank-deficient, full or sparse, numeric or symbolic, symmetric or unsymmetric, and real or complex. It efficiently handles triangular, permuted triangular, symmetric positive-definite, and Hessenberg matrices. When the matrix has specific known properties, the linsolve function can be faster (see Section 5.5, and a related Fortran code in Chapter 13).

21.3 Polynomial roots

Solving the function $f(x) = 0$ for the special case when f is a polynomial and x is a scalar is discussed in Section 20.1. The more general case is discussed in the next section.

21.4 Nonlinear equations

The fzero function finds a numerical solution to $f(x) = 0$ when f is a real function over the real domain (both x and $f(x)$ must be real scalars). This is useful when an analytic solution is not possible. You must supply either an initial guess, or two values of x for which the function differs in sign. Here is a simple example that computes $\sqrt{2}$.

```
fzero(@(x) x^2-2, 1)
```

The fzero function can only find an x for which $f(x)$ crosses the x-axis. If the sign of $f(x)$ does not differ on either side of x, the zero point x will not be found. Try this example. Create two anonymous functions (regular M-files can also be used):

```
fa = @(x) (x-2)^2
fb = @(x) (x-2)^2 - 1e-12
```

The zero of fa cannot be found, and neither can a zero of fb be found if your initial guess is too far from the solution. Both of these examples will fail:

```
fzero(fa, 1)
fzero(fb, 3)
```

Both functions can be easily solved with the Symbolic Math Toolbox. Note that solve correctly reports that 2 is a double root of (x-2)^2. Try:

```
syms x
solve((x-2)^2)
s = solve((x-2)^2-1e-12)
fb(s(1))
fb(s(2))
```

The zeros of fb can be found numerically only if you guess close enough, or if you provide two initial values of x for which fb differs in sign:

```
fzero(fb, 2)
format long
fzero(fb, [2 3])
fzero(fb, [1 2])
```

All of the functions used in the examples so far can be solved analytically. Here is one that cannot (also plot the function so that you can see where it crosses the *x*-axis):

```
f = @(x) real(airy(x))
figure(1) ; clf
ezplot(f)
solve('real(airy(x))')
```

The first zero is easy to compute numerically:

```
s = fzero(f, 0)
hold on
plot(s, f(s), 'ro')
```

The fminbnd function finds a local minimum of a function, given a fixed interval. This example looks for a minimum in the range -4 to 0.

```
xmin = fminbnd(f, -4, 0)
plot(xmin, f(xmin), 'ko')
```

To find a local maximum, simply find the minimum of -f.

```
```

```
g = @(x) -real(airy(x))
xmax = fminbnd(g, -5, -4)
plot(xmax, f(xmax), 'ko')
```

Now find the zero between these two values of x:

```
s = fzero(f, [xmax xmin])
plot(s, f(s), 'ro')
```

The fminbnd function can only find minima of real-valued functions of a real scalar. To find a local minimum of a scalar function of a real vector x, use fminsearch instead. It takes an initial guess for x rather than an interval. Try this example:

```
xmin = fminsearch(f, -6)
plot(xmin, f(xmin), 'kx')
```

21.5 Ordinary differential equations

The symbolic solution to the ordinary differential equation $y' = t^2 y$ appears in Section 19.12. Here is the same ODE, with a specific initial value of $y(0) = 1$, along with its symbolic solution.

```
syms t y
Y = dsolve('Dy = t^2*y', 'y(0)=1', 't')
```

Not all ODEs can be solved analytically, so MATLAB provides a suite of numerical methods. The primary method for initial value problems is ode45. For an ODE of the form $y' = f(t,y)$, the basic usage is:

```
[tt,yy] = ode45(@f, tspan, y0)
```

where @f is a handle for a function yprime=f(t,y) that computes the derivative of y, tspan is the time span to compute the solution (a 2-element vector), and y0 is the initial value of y. The variable t is a scalar, but y can be a

157

vector. The solution is a column vector `tt` and a matrix `yy`. At time `tt(i)` the numerical approximation to `y` is `yy(i,:)`.

To solve this ODE numerically, create an anonymous function:

```
f1 = @(t,y) t^2 * y
```

Now you can compute the numeric solution:

```
[tr,yr] = ode45(f1, [0 2], 1) ;
```

Compare it with the symbolic solution:

```
ts = 0:.05:2 ;
ys = subs(Y, t, ts) ;
figure(2) ; clf
plot(ts,ys, 'r-', tr,yr, 'bx') ;
legend('symbolic', 'numeric')
ys = subs(Y, t, tr) ;
[tr ys yr ys-yr]
err = norm(ys-yr) / norm(ys)
```

To solve higher-order ODEs, you need to convert your ODE into a first-order system of ODEs. Consider the ODE $y'' + y = t^2$ with initial values $y(0) = 1$ and $y'(1) = 0$. The symbolic solution to this ODE appears in Section 19.12, but here is the solution with initial values specified:

```
Y = dsolve('D2y + y = t^2', ...
    'y(0)=1', 'Dy(0)=0', 't')
```

Define $y_1 = y$ and $y_2 = y'$. The new system is $y'_2 = t^2 - y_1$ and $y'_1 = y_2$. Create an anonymous function:

```
f2 = @(t,y) [y(2) ; t^2-y(1)]
```

The function `f2` returns a 2-element column vector. The first entry is y'_1 and the second is y'_2. We can now solve this ODE numerically:

```
[tr,yy] = ode45(f2, [0 2], [1 0]') ;
yr = yy(:,1) ;
```

Note that `ode45` returns a 61-by-2 solution `yy`. Row `i` of `yy` contains the numerical approximation to y_1 and y_2 at time `tr(i)`. Compare the symbolic and numeric solutions using the same code for the previous ODE.

`ode45` can return a structure `s=ode45(...)` which can be used by `deval` to evaluate the numerical solution at any time `t` that you specify. There are seven other ODE solvers, able to handle stiff ODEs and for differential algebraic equations. Some can be more efficient, depending on the type of ODE you are trying to solve. Type `doc ode45` for more information.

21.6 Other differential equations

Delay differential equations (DDEs) are solved by `dde23`. The function `bvp4c` solves boundary value ODE problems. Finally, partial differential equations are solved with `pdepe` and `pdeval`. See the online help facility for more information on these ODE, DDE, and PDE solvers.

22 Displaying Results

The `format` command provides basic control over how your results are printed in the Command Window. Try this, for a trigonometric table with a few digits of precision:

```
format short
x = [0:.1:pi]' ;
f = {@sin, @cos, @tan, @cot} ;
y = x ;
for i = 1:length(f)
    y = [y f{i}(x)] ;
end
```

```
disp(y)
```

The `length` function gives the length of a vector or the largest dimension of an array (`length(f)` is 4 in this example). The cell array `f` is used in this example. Another way to construct `y` would be:

```
y = [x sin(x) cos(x) tan(x) cot(x)] ;
```

You can increase the number of digits printed with `format long`, but that does not allow you to define how many digits are printed. If you tried to add `pi/2` to the table, the `tan` column would contain a huge (erroneous) value that causes the rest of the digits in the table to be obscured. Try adding the statement `x=[x ; pi/2]` after `x` is first defined.

This problem is where `fprintf` is useful. If you know C, it acts just like the standard C `fprintf`, except that the reference to the file is optional in the MATLAB `fprintf`, and the MATLAB `fprintf` can print arrays. The basic syntax (like `printf` in C) is:

```
fprintf(format_string, arg1, arg2, ...)
```

The format string tells MATLAB how to print each argument (`arg1`, `arg2`, ...). It contains plain text, which is printed verbatim, plus special conversion codes that start with `'%'` (to print an argument) or `'\'` (to print a special character such as a newline, tab, or backslash). The basic syntax for a conversion code is `%W.Pc`, where `W` is the optional field width (the total number of characters used to represent the number), `P` is the optional precision (the number of digits to the right of the decimal point), and `c` is the conversion type. Both `W` and `P` are fixed integers. The dot before the `P` field is required only if `P` is specified. The most common conversion types are:

d   decimal (integer)

e   exponential notation (as in 2.3e+002)

f   fixed-point notation

g   e or f, whichever is more compact

s   string

Special characters include \n for newline, \t for tab, and
\\ for backslash itself. Here is a simple example that prints
π with 8 digits past the decimal point, in a space of 12
characters:

```
fprintf('pi is %12.8f\n', pi)
```

Try changing the 12 to 14, and you will see how fprintf
pads the string for pi to make it 14 characters wide. Note
the last character is '\n', which is a newline. If this were
excluded, the next line of output would start at the end of
this line. Sometimes that is what you want (see below for an
example).

Unlike printf or fprintf in the C language, the
MATLAB fprintf can print arrays. It accesses an array
column by column, and reuses the format string as needed.
This simple example prints the magic(3) array. It also
gives you an example of how to print a backslash and a
single quote:

```
A = magic(3)
fprintf('%4.2f %4.2f %4.2f\n', A')
b = (1:3)' ;
fprintf('A\\b is [%g %g %g]''\n', A\b);
```

The array A is transposed in the first fprintf, because
fprintf cycles through its data column by column, but
each use of the format string prints a single line of text as
one row of characters on the Command Window.
Fortunately it makes no difference for vectors:

```
fprintf('x is %d\n', 1:5)
fprintf('x is %d\n', (1:5)')
```

Here is a way of adding extra information to your display:

```
fprintf( ...
'row %d is %4.2f %4.2f %4.2f\n', ...
[(1:3)' A]')
```

Here is a revised trigonometric table using `fprintf` instead. A header has been added as well:

```
x = [0:.1:pi]' ;
f = {@sin, @cos, @tan, @cot} ;
y = x ;
fprintf('    x') ;
for i = 1:length(f)
    fprintf('    %s(x)',func2str(f{i}));
    y = [y f{i}(x)] ;
end
fprintf('\n') ;
fprintf( ...
'%3.2f %9.4f %9.4f %9.4f %9.4f\n',y');
```

`fprintf`, by default, prints to the Command Window. You can instead open a file, write to it with `fprintf`, and close the file. Add:

```
fid = fopen('mytable.txt', 'w') ;
```

to the beginning of the example. Add `fid` as the first argument to each `fprintf`. Finally, close the file at the end with the statement:

```
fclose(fid) ;
```

Your table is now in the file `mytable.txt`.

The `sprintf` function is just like `fprintf`, except that it sends its output to a string instead of the Command Window or a file. It is useful for plot titles and other annotation, as in:

```
title(sprintf('The result is %g', pi))
```

162

You cannot control the field width or precision with a variable as you can in the C printf or fprintf, but string concatenation along with sprintf or num2str can help here. Try:

```
for n = 1:16
    s = num2str(n) ;
    s = ['%2d digits: %.' s 'g\n'] ;
    fprintf(s, n, pi) ;
end
```

23 Cell Publishing

Cell publishing creates nicely formatted reports of MATLAB code, Command Window text output, figures, and graphics in HTML, LaTeX, PDF, or XML.

The term *cell publishing* has nothing to do with the cell array data type. In this context, a cell is a section of an M-file that corresponds to a section of your report. A cell starts with a cell divider, which is a comment with two percent signs at the beginning of a line, and ends either at the start of the next cell, or the end of the M-file. Cell publishing is normally done via scripts, not functions.

Create a new M-file with this 2-cell example. Give it the name myfirstcell.m.

```
%% Integrate a function
syms x
f = x^2
e = int(f)

%% Plot the results
figure(1)
ezplot(e)
```

Now publish the report to HTML, by selecting File ▶Publish myfirstcell.m or by clicking the publish button:

The M-file is evaluated and the report is presented in HTML form in a new window (the MATLAB Web Browser). The report is also saved to a file with the same name as your M-file but with an html file type. It includes the cell titles (the text after the double %%), the code itself, the output of the code, and any figures generated. You can change this default behavior in the File ▶Publish Configuration for myfirstcell.m menu option.

To run the M-file without publishing the results, simply click the run button, as usual, or select Cell ▶Evaluate Entire File. Individual cells can also be evaluated, one at a time. To publish and view the results from the MATLAB command line, use these commands:

```
publish myfirstcell.m
web html/myfirstcell.html
```

Additional descriptive text can be added as plain comments (one %) after the cell divider but before any commands. The text can be marked in various styles (bold, monospaced, TeX equations, and bullet lists, for example). See the Cell ▶Insert Text Markup ▶... menu for a complete list.

To add descriptive text without starting a new report section, start with a cell divider that has no title (a line containing just %%). This creates a new cell, but it appears in the same section of the report as the cell before it.

For a longer example of a report generated via cell publishing, download the factorize object discussed in Section 9.4, and look in the Factorize/html directory.

A Appendix: The MATLAB Top 500

Open the online Help window, select Help:MATLAB
►Functions and Help:Symbolic Math Toolbox ►Functions,
and you will see a lengthy index of all functions in
MATLAB and the Symbolic Math Toolbox. Many of these
functions are needed only in special cases, however, or for
very specific kinds of applications. This Appendix gives
you a more manageable index, with a description of the 500
most frequently used functions, operators, and special
characters in MATLAB and the Symbolic Math Toolbox, in
the same outline as the Help ►Functions documentation. In
this Appendix, a page reference in brackets refers to a
discussion of the function in the main part of the book.
Other page references refer to this Appendix.

B Desktop Tools and Development Environment

B.1 Command Window and History

`clc` clears the Command Window display, giving you an empty
screen. [p. 6]

`diary` logs Command Window output to a file. diary on starts
the log; diary off stops it. Flush the diary by turning it off and
then on again. diary(s) uses the filename given by the string s.
Default filename is diary. [p. 5]

`format` changes how numeric values are displayed in the
Command Window. Try format long and format short. Has no
effect on fprintf. [p. 6]

`system` executes a command in your operating system shell.
Try system('ls') in UNIX. See also p. 191 for the ! character.
[p. 103]

B.2 Help for Using MATLAB

`help` prints the help text for a function in the Command Window.
Try help disp, or help *function* for any other function. [p. 42]

`web` opens a web browser. Try web www.mathworks.com. Also
useful for viewing HTML files created via cell publishing. [p. 164]

B.3 Workspace

`clear` deletes variables from workspace. clear, by itself, clears
all variables. clear x y clears just x and y. clear classes clears
object class definitions. [p. 10]

`delete` deletes a file or graphics object. [p. 8]

`exist` determines if a variable, file, folder, or class exists.
exist('x') returns a value from 0 to 8, depending on what x is.
A second parameter ('file', 'var', 'class', 'builtin', or
'dir') only looks for items of that type, and returns 1 if it exists, 0
otherwise. [p. 105]

`which` determines which function or file a word refers to, or
whether there is a variable with that name. Try i=3;
which i -all. [p. 71]

`whos` lists variables in the workspace. whos prints a list.
s=whos returns a struct with information about each variable. [p. 9]

B.4 Managing Files

B.4.1 Search Path

`addpath` adds folders to the search path. Try addpath(pwd).

`path` displays or changes the search path. [p. 43]

`which` determines what function or file a name refers to. See p. 166.

B.4.2 File Operations

`cd` changes the current folder (short for *change directory*). [p. 8]

`copyfile` makes a copy of a file.

`delete` deletes a file. [p. 8]

`dir` lists the files and folders in a folder. `dir` by itself lists the current folder. `dir('F')` or `dir F` lists the folder F. [p. 8]

`isdir` tests whether a string refers to a folder.

`ls` lists the contents of the current folder.

`matlabroot` lists the folder where MATLAB is installed.

`mkdir` creates a folder.

`pwd` returns the current folder as a string. [p. 8]

`type` displays the contents of a file. Try `type factorial`. [p. 8]

B.5 Programming Tools

`computer` returns information about your computer.

`system` executes an operating system command. See p. 166.

`ispc` tests if your computer is running Microsoft Windows.

`ver` displays the version of MATLAB and all Toolboxes.

`version` returns the MATLAB version number as a string.

C Data Import and Export

C.1 File Name Construction

`fileparts` extracts the components of a filename and its path.

`filesep` returns the file separator ('\' on Microsoft Windows, '/' on Linux or Mac).

`fullfile` constructs the full filename (with its entire path) from a list of folders and a base filename.

C.2 File Opening, Loading, and Saving

`load` loads variables from a MAT-file. [p. 10]

`open` opens a file in the Editor, Workspace, or Figure Window.

`save` saves variables to a MAT-file. [p. 10]

`uigetdir` displays a dialog box for selecting a directory.

`uigetfile` displays a dialog box for accessing files.

`uiputfile` displays a dialog box for saving files.

C.3 Low-Level File I/O

`fclose` closes a file opened by fopen. [p. 162]

`feof` tests if the end-of-file has been reached.

`fgetl` reads a line from a text file, discarding newline characters.

`fgets` reads a line from a text file, keeping newline characters.

`fopen` opens a file. f=fopen('foo','w') opens the file 'foo' for writing, creating it if necessary. f=fopen('foo','r') opens the file 'foo' for reading. [p. 162]

fprintf displays numbers and strings (or prints them to a file) with tight control over how they are displayed. Try fprintf('%10.1e\n',eps). [p. 160]

fread reads data from a binary file. With the file from the fwrite example below, try f=fopen('a.bin','r'); C=fread(f,'double');, and compare with A(:).

fscanf reads formatted data from a text file, with a format string similar to the one used by fprintf.

fseek moves to a specified position in a file, where the next fread or fwrite will take place.

ftell returns the current position in a file, as the number of bytes from the beginning of the file.

fwrite writes binary data from a file. Try A=rand(4); f=fopen('a.bin','w'); fwrite(f,A,'double'); fclose(f). Then try the fread example above.

C.4 Text Files

textscan reads formatted data from a text file or string.

type displays the contents of a file. Try type factorial. [p. 8]

C.5 Audio and Video

wavread reads an audio signal from a .wav file.

sound plays audio from a signal. Try load handel; sound(y,Fs).

C.6 Images

imread reads an image from a graphics file. Supported formats include TIFF, JPEG, GIF, PNG, BMP, ICO, and may others. Image processing typically uses uint8, uint16, and other compact integer data types. Use image to display an image.

imwrite writes an image to a graphics file.

169

D Mathematics

D.1 Arrays and Matrices

D.1.1 Basic Information

`disp` displays a MATLAB variable in the Command Window. This works for all data types in MATLAB (scalars, vectors, matrices, higher-dimensional arrays, strings, cell arrays, structs, and objects). Use `format` to control the detail displayed by `disp`. Use `fprintf` instead of `disp` for even more control. [p. 18]

`display` displays a MATLAB variable or expression in the Command Window, as if it was the result of a statement with no trailing semicolon.

`isempty` determines if a matrix has no entries. `isempty(A)` is true if A is an empty array with one or more dimension of zero size. Same as `min(size(A))==0`. [p. 23]

`isequal` tests two variables for equality, returning 1 if equal, 0 otherwise. `isequal(a,b)` compares two scalars, arrays, structs, cells, objects, or any other pair of MATLAB variables. [p. 52]

`isfinite` tests if a number is finite (not `+inf`, `-inf`, or `nan`).

`isinf` tests if a number is `+inf` or `-inf`.

`islogical` tests if a variable is `logical`. See p. 188.

`isnan` tests if a number is a `NaN`.

`isnumeric` tests if a variable has a numeric type. See p. 188.

`isscalar` tests if a variable is a scalar.

`isvector` tests if a variable is a row or column vector. A vector has size n-by-1 or 1-by-n with n>=0.

`length` returns the length of an array. `length(A)` is the number of entries along the largest dimension of A. It is `max(size(A))` if A is non-empty, or zero otherwise. [p. 160]

`max` finds the largest entries in an array. `max(x)` for a vector x returns a scalar. `max(A)` for a matrix returns a row vector of length `size(A,2)` with the largest entry in each column. For computing the maximum along other dimensions, use `max(A,[],d)` for dimension d. `C=max(A,B)` returns an array the same size as A and B, where each entry in C is the larger of the two corresponding entries in A and B. A second output returns the indices of the largest entries in A. [p. 25]

`min` finds the smallest entries in an array. The syntax of min is the same as max, above. [p. 33]

`ndims` returns the number of array dimensions. `ndims(x)` is always ≥ 2. `ndims(rand(5,5,5))` is 3, for a 3-D array.

`numel` number of elements in an array. `numel(A)` is the same as `prod(size(A))`.

`size` returns the size of an array. `d=size(A)` returns d as a vector of length `ndims(A)` (normally of size 2), with the size of each dimension of A. Try `size(rand(3,4))`. `[m,n]=size(A)` returns the dimensions as scalars m and n. `size(A,2)` is the number of columns of a 2-D array A. [p. 23]

D.1.2 Operators

See 190. [See also pp. 14–16 for a discussion of matrix and entry-wise operators.]

D.1.3 Elementary Matrices and Arrays

`diag` extracts the diagonal of a matrix, or creates a diagonal matrix. `diag(A)` is the diagonal of a matrix A. `diag(A,k)` is the kth diagonal (use k=-1 for the subdiagonal). `diag(x)` for a vector x is a diagonal matrix. [p. 23]

`eye` returns an identity matrix. `eye(n)` is the n-by-n identity matrix. `eye(m,n)` or `eye([m n])` is an m-by-n matrix with ones on the diagonal and zeros elsewhere. [p. 24]

`ind2sub` converts linear indices to subscripts.

linspace creates a linearly spaced vector. linspace(a,b) is a row vector of 100 points between a and b. linspace(a,b,n) generates n points. [p. 20]

logspace creates a logarithmically spaced vector. logspace(...) is 10.^linspace(...), except that the default number of points is 50, not 100.

meshgrid creates X, Y, and Z arrays for 3-D plots. [p. 114]

ndgrid creates X, Y, and Z arrays for plots of N-D functions.

rand computes uniformly distributed pseudo-random numbers. rand(n) is an n-by-n array of random numbers. rand(m,n) is m-by-n. [p. 23]

randn computes normally distributed pseudo-random numbers. Same syntax as rand. [p. 23]

sub2ind converts subscripts to linear indices. If k=find(A) and [i,j]=find(A), the k=sub2ind(size(A),i,j).

ones returns a matrix of all ones. Same syntax as zeros, below. [p. 23]

zeros returns a matrix of all zeros. zeros(n) is an all-zero n-by-n matrix. zeros(m,n) is of size m-by-n. The inputs can also be given as a vector whose length is the dimension of the array. For higher dimensional arrays, simply use more input parameters. [p. 23]

D.1.4 Array Operations

bsxfun applies a binary function with singleton expansion. Try C=bsxfun(@minus, A, mean(A)); sum(C). [p. 32]

cross computes the cross product of two vectors of length 3.

cumprod computes the cumulative product. If y=cumprod(x), then y(k) is prod(x(1:k)).

cumsum computes the cumulative sum. If y=cumsum(x), then y(k) is sum(x(1:k)).

dot computes the dot product. For two column vectors, dot(x,y) is x'*y.

kron computes the Kronecker tensor product.

prod computes the products of the entries of an array. For a vector, prod(x) is the product of the entries. For a matrix, prod(A) is a row vector of the product of each column of A. prod(A,2) is a column vector of row products. [p. 25]

sum sums entries along one dimension. For a vector, sum(x) is the sum of the entries. For a matrix, sum(A) is a row vector of the sums of each column of A. sum(A,2) is a column vector of row sums. [p. 46]

tril extracts the lower triangular part of a matrix. tril(A) is the lower triangular part. tril(A,k) is all zero above the kth diagonal and equal to A elsewhere. [p. 121]

triu extracts the upper triangular part of a matrix. triu(A) is the upper triangular part. triu(A,k) is all zero below the kth diagonal and equal to A elsewhere. [p. 121]

D.1.5 Array Manipulation

cat concatenates arrays along a given dimension. cat(1,A,B) is the same as [A;B] and cat(2,A,B) is [A,B]. [p. 57]

circshift shifts the entries of a matrix.

diag extracts the diagonal of a matrix, or creates a diagonal matrix. See p. 171.

end specifies the end of an array dimension. A(k,3:end) is the kth row of A, excluding the first two entries. [p. 22]

fliplr flips a matrix along its columns. fliplr(A) is the same as A(:,end:-1:1).

flipud flips a matrix along its rows. flipud(A) is the same as A(end:-1:1,:).

permute rearranges the dimensions of an N-D array. permute(A,[2 1]) for a 2-D array is the same as A.'.

repmat replicates and tiles an array. C=repmat(A,m,n) creates a matrix C by replicating A, m times along the rows of C, and n times along the columns. [p. 32]

`reshape` reshapes an array. `C=reshape(A,m,n)` is m-by-n with entries taken column-wise from A. [p. 30]

`rot90` rotates a matrix, like rotating an image.

`sort` sorts entries in an array. `sort(x)` sorts a row or column vector x. `sort(A)` sorts each column of A. `sort(A,2)` sorts each row. `[C,i]=sort(...)` returns the permutation i from the sort. [p. 60]

`sortrows` sorts the rows of a matrix. `sortrows(S)` for a character array is a dictionary sort. `sortrows(A)` for a numeric matrix sorts A in ascending order according to the first column, with ties broken by subsequent columns. A second argument changes which columns to use, and whether to sort in ascending or descending order.

`squeeze` removes singleton dimensions of N-D arrays. `squeeze(rand(3,1,4))` is a 3-by-4 matrix. Has no effect on 2-D arrays.

D.2 Linear Algebra

D.2.1 Matrix Analysis

`cond` computes the condition number of a matrix. `cond(A)` is the 2-norm condition number, $||A||_2||A^{-1}||_2$, or the ratio of the largest singular value over the smallest. `cond(A,p)` is $||A||_p||A^{-1}||_p$ for p=1, 2, `'fro'`, or `inf`. [p. 61]

`det` computes the determinant of a matrix. Never use `det` to test for singularity; use `rank(A) < min(size(A))` instead.

`norm` computes the norm of a matrix. `norm(A)` is $||A||_2$, `norm(A,p)` is $||A||_p$ for p=1, 2, `inf`, or `'fro'` (for Frobenius norm). [p. 68]

`null` computes a basis for the null space of a matrix.

`rank` computes the rank of a matrix. `rank(A)` is the number of linearly-independent rows or columns of A, equivalently the number of singular values of A that are not too tiny.

`trace` sums the diagonal entries of a matrix, `trace(A)` is `sum(diag(A))`.

D.2.2 Linear Equations

`chol` computes the Cholesky factorization of a symmetric positive definite matrix. R=chol(A) is upper triangular, so that R'*R=A. [R,k,P]=chol(A) returns a positive integer k and a permutation matrix P so that R'*R=P'*A*P. If k<size(A,1) then A is not positive definite. P is available for sparse A only; using it leads to a much sparser R. With a second input, 'vector', P is a permutation vector. [p. 27]

`cond` computes the condition number of a matrix. See p. 174.

`inv` computes A^{-1}. This is the most frequently abused function in MATLAB. Never compute inv(A)*B or B*inv(A). Instead, use A\B or B/A, respectively. To use something like the inv(A)*B syntax, download the factorize package from the MATLAB File Exchange or from the web page for this book. With that package, inverse(A)*B computes inv(A)*B without actually computing the inverse. It uses a chol, lu, or qr factorization of A instead, and solves A\B. In the rare case when you actually need a few entries of the inverse, use S=inverse(A) and S(1,1) (for example).

`lu` computes the LU factorization. [L,U,P]=lu(A) returns a lower triangular matrix L, an upper triangular matrix U, and a permutation matrix P so that L*U=P*A. A fourth output Q is available when A is sparse; using it leads to much sparser factors L and U. With two outputs [L,U]=lu(A), L is a permuted triangular matrix. With a second input, 'vector', P and Q are permutation vectors. [pp. 58, 124]

`pinv` computes the Moore-Penrose pseudo-inverse.

`qr` computes the QR factorization. [Q,R]=qr(A) returns a unitary matrix Q and an upper triangular matrix R so that Q*R=A. [Q,R,E]=qr(A) computes Q*R=A*E instead. If A is full, E is chosen so that abs(diag(R)) is monotonically decreasing. If A is sparse, E is chosen to reduce the number of nonzeros in R. R=qr(A) returns only R.

D.2.3 Eigenvalues and Singular Values

`eig` computes the eigenvalues and eigenvectors of a matrix, or finds the generalized eigenvalues/eigenvectors. d=eig(A) is a vector of eigenvalues. [V,D]=eig(A) gives A*V=V*D where D is diagonal. [V,D]=eig(A,B) gives A*V=B*V*D. For the sparse case, can only to find eigenvalues for a sparse symmetric A. [p. 26]

poly constructs a polynomial from a set of specified roots, or constructs the characteristic polynomial of a matrix. [p. 150]

svd computes the singular value decomposition.
[U,S,V]=svd(A) computes two unitary matrices U and V (with left and right singular vectors) and a diagonal matrix S (containing the singular values) so that U*S*V'=A. The rank, cond, pinv, and null functions rely on svd. Does not work for sparse A.

D.2.4 Factorization

chol computes the Cholesky factorization. See p. 175.

lu computes the LU factorization. See p. 175.

qr computes the QR factorization. See p. 175.

svd computes the singular value decomposition. See above.

D.3 Elementary Math

D.3.1 Trigonometric

The 6 basic trigonometric functions operate entry-wise on each element of an array. [p. 24]

sin computes the sine in radians.

cos computes the cosine in radians.

tan computes the tangent in radians. $\tan(x) = \sin(x)/\cos(x)$.

sec computes the secant in radians. $\sec(x) = 1/\cos(x)$.

csc computes the cosecant in radians. $\csc(x) = 1/\sin(x)$.

cot computes the cotangent in radians. $\cot(x) = \cos(x)/\sin(x)$.

Each of the 6 functions above has 6 different variants. Append d for degrees instead of radians. Append h for the hyperbolic version. Prepend a for the inverse. All 36 functions are listed below.

sin	sind	sinh	asin	asind	asinh
cos	cosd	cosh	acos	acosd	acosh
tan	tand	tanh	atan	atand	atanh
sec	secd	sech	asec	asecd	asech
csc	cscd	csch	acsc	acscd	acsch
cot	cotd	coth	acot	acotd	acoth

$\boxed{\texttt{atan2}}$ is the four-quadrant version of atan. Use atan2(y,x) instead of atan(y/x) to compute $\tan^{-1}(y/x)$.

D.3.2 Exponential

$\boxed{\texttt{exp}}$ computes the exponential of entries in an array. exp(x) is e^x for the scalar x. [p. 106]

$\boxed{\texttt{log}}$ computes the exponential of entries in an array. log(x) is $\log x$ for the scalar x. [p. 153]

$\boxed{\texttt{log10}}$ computes the base-10 logarithm. [p. 24]

$\boxed{\texttt{log2}}$ computes the base-2 logarithm, and dissects a number x into its mantissa f and integer exponent e, so that x=f*2^e. [p. 46]

$\boxed{\texttt{sqrt}}$ sqrt(A) is the square root of each entry of A. [p. 24]

D.3.3 Complex

$\boxed{\texttt{abs}}$ computes the absolute value of each entry in an array. [p. 34]

$\boxed{\texttt{angle}}$ computes the phase angle of each entry in an array.

$\boxed{\texttt{complex}}$ constructs a complex number from two real numbers. complex(A,B) is A+1i*B.

$\boxed{\texttt{conj}}$ computes the complex conjugate of each entry in an array.

$\boxed{\texttt{i}}$ returns the imaginary unit, sqrt(-1). Often shadowed by the commonly used loop index i. Use 1i instead. See also j. [p. 18]

$\boxed{\texttt{imag}}$ returns the imaginary part of a number. [p. 63]

$\boxed{\texttt{isreal}}$ tests if a variable is real.

`real` returns the real part of a complex number. [p. 119]

`sign` returns the sign of a number. `sign(x)` is 1 if x>0, zero if x==0, and -1 if x<0. [p. 39]

D.3.4 Rounding and Remainder

`ceil` rounds towards $+\infty$. `ceil(1.7)` is 2 and `ceil(-1.7)` is -1. [p. 24]

`fix` rounds towards zero. `C=fix(A)` rounds each entry in A. `fix(1.7)` is 1 and `fix(-1.7)` is -1. The term dates back to the early days of Fortran, which has a function `ifix` that rounds a floating-point number to a *fix*ed-point integer. [p. 20]

`floor` rounds towards $-\infty$. `floor(1.7)` is 1 and `floor(-1.7)` is -2. [p. 24]

`mod` computes the modulus after division. `m=mod(x,y)` is the remainder after the integer division x/y, for the integer division conventionally used in discrete mathematics. [p. 47]

`rem` computes the remainder after division. `r=rem(x,y)` is the remainder after the integer division x/y, for the integer division conventionally used in computer programming languages. [p. 24]

`round` rounds towards the nearest integer. [p. 24]

D.3.5 Discrete Math

`factor` computes the prime factors of a nonnegative integer. Try `factor(42)`, which is `[2 3 7]`. `prod(factor(n))` is n. [p. 136]

`factorial` computes the factorial. `factorial(n)` is $n!$, or `prod(1:n)`.

D.4 Polynomials

`conv` performs convolution and polynomial multiplication. Try `conv([1 3 2],[3 -2 1])`, which gives `[3 7 1 -1 2])`, since $(x^2 + 3x + 2)(3x^2 - 2x + 1) = 3x^4 + 7x^3 + x^2 - x + 2$. [p. 152]

poly constructs a polynomial from a set of specified roots, or constructs the characteristic polynomial of a matrix. Try poly([1 2]), which creates the polynomial $(x-1)(x-2) = x^2 - 3x + 2$, and returns the coefficients [1 -3 2]). [p. 150]

polyfit fits a polynomial to data. [p. 151]

polyval evaluates a polynomial. Try polyval([1 -3 2], 7), which evaluates $p(7)$ where $p(x) = x^2 - 3x + 2$. [p. 151]

roots computes the roots of a polynomial. Try roots([1 -3 2], 7), which returns the roots [2;1]. [p. 150]

D.5 Interpolation and Computational Geometry

D.5.1 Interpolation

interp1 interpolates a 1-D function. Try x=0:.1:2*pi; y=sin(x); interp1(x,y,pi) for an interpolation of $\sin(\pi)$ using a coarse grid.

meshgrid creates X, Y, and Z arrays for 3-D plots. [p. 114]

ndgrid creates X, Y, and Z arrays for plots of N-D functions.

spline constructs a cubic spline interpolation. [p. 152]

D.5.2 Domain Generation

meshgrid creates X, Y, and Z arrays for 3-D plots. [p. 114]

ndgrid creates X, Y, and Z arrays for plots of N-D functions.

D.6 Nonlinear Numerical Methods

D.6.1 Ordinary Differential Equations

ode45 is the primary ODE solver in MATLAB. [p. 157]

odeset creates an options struct for ODE solvers.

D.6.2 Optimization

fminsearch finds the minimum of a function. Try
fminsearch(@cos,3), which finds a 6-digit approximation of
pi. [p. 157]

fzero finds a root of a continuous function. Try
fzero(@sin,3), which finds pi. [p. 155]

optimset creates an options struct for optimization functions.

D.6.3 Numerical Integration (Quadrature)

quad performs numerical integration. Try quad(@sin,0,pi),
and compare with syms x; int(sin(x),0,pi). [p. 153]

D.7 Specialized Math

besselj computes the Bessel function of the first kind, $J_v(z)$.

beta computes the Beta function, $B(x, y)$.

gamma computes the Gamma function, $\Gamma(a)$. For positive
integers, $\Gamma(n) = (n-1)!$.

psi computes the psi (polygamma) function, $\psi_k(x)$.

D.8 Sparse Matrices

find finds indices and values of nonzeros. i=find(A) returns
1-D indices of all nonzeros in a full or sparse matrix A.
i=find(A,k) returns just the first k entries; add the string 'last'
as a third argument to get the last k. [i,j,x]=find(A, ...)
returns 2-D indices and values x, so that x(k)=A(i(k),j(k)) is
the kth nonzero in A. [p. 28]

full converts a matrix to a full representation. [p. 120]

sparse builds a sparse matrix from a list of entries (via S=sparse(i,j,x)) or converts a full matrix to a sparse representation (via S=sparse(A)). sparse(m,n) is an all-zero m-by-n sparse matrix. You should normally not create a sparse matrix S=sparse(A) from a full matrix A. Use S=sparse(i,j,x) instead, which builds a sparse matrix from three vectors containing a list of nonzero entries. [p. 122]

nnz finds the number of nonzeros in a matrix (whether sparse or full). [p. 120]

randperm constructs a pseudo-random permutation. randperm(n) is a permutation of 1:n.

D.9 Math Constants

eps determines the floating-point relative accuracy. eps is the smallest number so that 1+eps>eps. eps(x) is the distance to the next floating-point number larger than x. On most computers, eps is 2^{-52}), or about 2.2×10^{-16}. [p. 39]

i returns the imaginary unit, sqrt(-1). Often shadowed by the commonly used loop index i. Use 1i instead. See also j. [p. 18]

inf is the IEEE representation of ∞. Also appears as Inf. The syntax of inf is the same as zeros, ones, and nan (for creating a matrix of Inf's). [p. 132]

nan is the IEEE representation of Not-a-Number (the result of dividing zero by zero, for example). Also appears as NaN. The syntax of nan is the same as zeros, ones, and inf (for creating a matrix of NaN's). NaN's propagate, since any computation with a NaN produces a NaN. Comparisons with NaN always return 0 (false). Thus, x==x is true for all numbers except for x=nan.

pi is a 16-digit approximation to π. [p. 6]

E Data Analysis

E.1 Basic Operations

`cumprod` computes the cumulative product. See p. 172.

`cumsum` computes the cumulative sum. See p. 172.

`prod` computes the products of the entries of an array. See p. 173.

`sort` sorts entries in an array. See p. 174.

`sortrows` sorts the rows of a matrix. See p. 174.

`sum` sums entries along one dimension. See p. 173.

E.2 Descriptive Statistics

`cov` computes the covariance matrix, or the variance of a vector.

`max` finds the largest entry in an array. See p. 171.

`mean` computes the mean. mean(x) is the mean of a vector x. mean(A) for a matrix is a row vector with the mean of each column of A. mean(A,2) is the mean of each row. [p. 25]

`median` computes the median. median(x) is the median of a x. median(A) for a matrix is a row vector with the median of each column of A. median(A,2) is the median of each row. [p. 25]

`min` finds the smallest entry in an array. See p. 171.

`mode` computes the mode. mode(x) is the mode of a vector x. mode(A) for a matrix is a row vector with the mode of each column of A. mode(A,2) is the mode of each row. [p. 25]

`std` computes the standard deviation. std(x) is the standard deviation of a vector x. std(A) for a matrix is a row vector with the standard variation of each column of A. [p. 25]

`var` computes the variance. var(x) is the variance of a vector x. var(A) for a matrix is a row vector with the variance of each column of A. [p. 25]

E.3 Filtering and Convolution

`conv` performs convolution and polynomial multiplication. See p. 178.

`conv2` performs 2-D convolution.

`filter` applies a digital filter to a data sequence.

E.4 Interpolation and Regression

`interp1` interpolates a 1-D function. See p. 179.

`polyfit` fits a polynomial to data. [p. 151]

`polyval` evaluates a polynomial. [p. 151]

E.5 Fourier Transforms

`abs` computes the absolute value of each entry in an array. [p. 34]

`angle` computes the phase angle of each entry in an array.

`fft` computes the discrete Fourier transform (DFT). `fft(x)` is the DFT of the vector x. `fft(A)` for a matrix computes the DFT of each column.

`fft2` computes the 2-D DFT.

`fftshift` centers the spectrum by shifting the zero-frequency component.

`ifft` computes the inverse DFT. If `y=fft(x)` for a vector x, the `x=ifft(y)`.

`ifft2` computes the inverse 2-D DFT.

E.6 Derivatives and Integrals

`diff` computes differences between adjacent entries. `diff((1:5).^2)` is `[3 5 7 9]`. [p. 129]

`gradient` computes the numerical gradient.

183

F Programming and Data Types

F.1 Data Types

F.1.1 Numeric Types

`cat` concatenate arrays. See p. 173.

`class` returns the class of a variable as a string.
class(eye(2)) is 'double', class(@log) is
'function_handle', and class('s') is 'char', for example.

`find` returns the indices and values of nonzeros in a matrix. See
p. 180.

`isa` tests whether a variable is from a given class
(isa(x,'double'), for example). See p. 188.

`isequal` tests two variables for equality. See p. 170.

`isfinite` tests if a number is finite (not +inf, -inf, or nan).

`isinf` tests if a number is +inf or -inf.

`isnan` tests if a number is a NaN.

`isnumeric` tests if a variable has a numeric type. See p. 188.

`isreal` tests if a variable is real (as opposed to complex).

`isscalar` tests if a variable is a scalar.

`isvector` tests if a variable is a vector. See p. 170.

`permute` rearranges the dimensions of an N-D array. See p. 173.

`reshape` reshapes an array. See p. 174.

`squeeze` removes singleton dimensions. See p. 174.

`zeros` returns an array of all zeros. See p. 172.

F.1.2 Characters and Strings

`cellstr` creates a cell array of strings from a character array. If S is m-by-n, then C=cellstr(S) is a cell array of length m, with each entry a row of S.

`char` converts an array of integers to a character array. [p. 101]

`eval` executes MATLAB expressions or statements from a string. See p. 198.

`findstr` Finds strings within another string. Use strfind instead.

`regexp` searches a string for a *regular expression*, which is a kind string that can match many strings. For example, to search for a digit followed by a colon, try s=datestr(now), regexp(s,'[0-9]:').

`sprintf` creates a formatted string of other numbers and strings, with tight control over how they are displayed. [p. 162]

`sscanf` reads formatted data from a string, with a format string similar to the one used by fprintf.

`strcat` concatenates strings horizontally. See p. 192.

`strcmp` compares strings. See p. 193.

`strcmpi` is a case-insensitive version of strcmp. See p. 193.

`strfind` determines if one string is a substring of another string. Try strfind('look in this book','oo'), which gives [2 15].

`strmatch` determines if one string is a prefix of another. See p. 194.

`strrep` finds and replaces substrings. strrep(s,a,b) replaces all occurrences of the substring a with b in the string s.

`strtrim` removes leading and trailing spaces.

F.1.3 Structures

class determines the class of a variable. See p. 184.

deal distributes inputs to outputs. Useful for assigning fields of a struct array from a cell array. If `s` is a 1-by-4 struct array, try `c={1,2,3,4}; [s.f]=deal(c{:})`, which is the same as `for i=1:4, s(i).f=c{i}; end`.

fieldnames returns a cell array of strings with the name of each member of a struct or object.

getfield gets one or more fields from a struct. `getfield(x,'y')` is the same as `x.y`.

isa tests whether a variable is from a given class (`isa(s,'struct')`, for example). See p. 188.

isequal tests two structs for equality. The order of members of each struct does not matter. See p. 170.

isfield checks for a member of a struct. `isfield(x,'s')` is true if `x.s` is a member of the struct `x`. Use `isfield(x,c)` where `c` is a cell array to check for multiple fields.

isscalar tests if a variable is a scalar. A 1-by-1 struct is considered scalar, even if it contains multiple members or array members.

isstruct tests if a variable is a struct. See p. 188.

isvector tests if a variable is a vector. See p. 170.

rmfield removes fields from a struct.

setfield sets fields in a struct.

struct creates a struct array. Try `x.a=1; x.b='s'`, which can also be done with `x=struct('a',1,'b','s')`. [p. 54]

struct2cell converts a struct to a cell array. See p. 187.

F.1.4 Cell Arrays

`cell` constructs an empty cell array. `cell(m,n)` is a m-by-n cell array, with every entry equal to the empty array (`[]`). [p. 53]

`cell2mat` constructs a matrix from a cell array of matrices. See p. 190.

`cellfun` applies a function to each entry of a cell array.

`cellstr` creates a cell array of strings from a character array. See p. 185.

`class` determines the class of a variable. See p. 184.

`deal` distributes inputs to outputs. See p. 186.

`isa` tests whether a variable is from a given class (`isa(c,'cell')`, for example). See p. 188.

`iscell` tests if a variable is a cell array. See p. 188.

`iscellstr` tests if a variable is a cell array of strings.

`isequal` tests two cell arrays for equality. See p. 170.

`isscalar` tests if a variable is a scalar. A 1-by-1 cell array is a scalar, even if it contains an array. Try `isscalar({rand(4)})`.

`isvector` tests if a variable is a vector. See p. 170.

`num2cell` converts a numeric array into a cell array, placing each entry in its own cell. See p. 190.

`struct2cell` converts a struct to a cell array, placing each field in its own cell.

F.1.5 Function Handles

`class` determines the class of a variable. See p. 184.

`feval` evaluates a function via a function handle. `feval(@f,x)` is the same as `f(x)` for the function handle `@f`. `feval` remains in MATLAB for historical reasons.

`isa` tests whether a variable is from a given class (isa(f,'function_handle'), for example). See p. 188.

`isequal` tests two function handles for equality. See p. 170.

`str2func` constructs a function handle from a string. str2func('f') is the same as @f. [p. 68]

F.1.6 Data Type Identification

`isa` tests whether a variable is from a given class. isa(x,'struct') is true if x is a struct, for example. Classes include logical, char, numeric (integer or floating-point), integer, float, int8 (and variants), single, double, cell, struct, function_handle, and any MATLAB or Java class.

`iscell` tests if a variable is a cell array. Same as isa(x,'cell').

`iscellstr` tests if a variable is a cell array of strings.

`ischar` tests if a variable is a string (a char array). Same as isa(x,'char'). [p. 70]

`isfield` checks for a member of a struct. See p. 186.

`islogical` tests if a variable is logical. Same as isa(x,'logical').

`isnumeric` tests if a variable has a numeric type (integer or floating-point). Same as isa(x,'numeric').

`isreal` tests if a variable is real (as opposed to complex).

`isstruct` tests if a variable is a struct. Same as isa(x,'struct').

`whos` lists variables in the workspace. See p. 166.

F.2 Data Type Conversion

F.2.1 Numeric

`double` converts numbers to IEEE double precision. [p. 56]

`int16` converts numbers to 16-bit signed integers.

`int32` converts numbers to 32-bit signed integers.

`single` converts numbers to IEEE single precision.
Computations in `single` are less accurate than `double`, but can be faster. [p. 56]

`uint8` converts numbers to 8-bit unsigned integers. Often used for images. [p. 56]

`uint16` converts numbers to 16-bit unsigned integers.

F.2.2 String to Numeric

`hex2dec` converts a string with a hexadecimal number to a number. hex2dec('1F') is 31.

`str2double` converts a string to a single number.
str2double('3.14') is the number 3.14. str2double(c) with a cell array of strings converts each entry in c. [p. 19]

`str2num` converts a string to one or more numbers, via eval. Use str2double to avoid side-effects of functions in the string. [p. 19]

F.2.3 Numeric to String

`char` converts an array of integers to a character array. [p. 101]

`dec2bin` converts a number to a string, in binary.

`int2str` converts an integer to a formatted string, for display. Try int2str(magic(5)).

`mat2str` converts an integer to a formatted string, for display or evaluation. Try mat2str(magic(5)).

`num2str` converts a numeric array to a string, for display. num2str(A) uses the default format %11.4g (a width of 11, with 4 digits displayed). num2str(pi,8) displays 8 digits of π, and num2str(A,s) uses the format string s (like fprintf). [p. 163]

F.2.4 Other Conversions

`cell2mat` constructs a matrix from a cell array of matrices. Try
A=magic(5) and C=num2cell(A). Then cell2mat(C)
reconstructs A. For the 2-D case, the matrices in each row of the
cell array (C{i,:}) must all have the same number of rows, and the
matrices in each column (C{:,j}) must all have the same number
of columns.

`datestr` converts the result from clock or now into a string.
See p. 196.

`logical` converts an array to logical. See p. 195.

`num2cell` converts a numeric array into a cell array, placing
each entry in its own cell. Try A=magic(5); C=num2cell(A),
which creates a cell array C so that C{i,j}=A(i,j).

`str2func` constructs a function handle from a string. See
p. 188.

`struct2cell` converts a struct to a cell array. See p. 187.

F.3 Operators and Special Characters

[See pp. 14–16, 33, 123, and 142 for a discussion of each operator.]

F.3.1 Arithmetic Operators

A+B	adds two matrices, or a matrix and a scalar
A-B	subtracts two matrices, or a matrix and a scalar
-A	negates the matrix A
A*B	multiplies two matrices, or a matrix and a scalar
B/A	solves $XA = B$. Use in place of B*inv(A)
A\B	solves $AX = B$. Use in place of inv(A)*B
A^B	A to the power B. A or B must be a scalar
A.*B	element-wise array multiplication
A./B	element-wise array right division
A.\B	element-wise array left division
A.^B	element-wise array power
A'	matrix transpose (complex conjugate, if complex)
A.'	array transpose (not complex conjugate)

F.3.2 Relational Operators

A<B	less than
A<=B	less than or equal to
A>B	greater than
A=>B	greater than or equal to
A==B	equal to
A~=B	not equal to

F.3.3 Logical Operators

A&&B	short-circuit logical *and* (scalars only)
A||B	short-circuit logical *or* (scalars only)
A&B	logical *and*
A|B	logical *or*
~A	logical *not*

F.3.4 Special Characters

=	assignment	[p. 3]
:	colon operator	[pp. 19–22]
( )	input parameters, precedence, subscript	[pp. 5, 13]
[ ]	output parameters, construct array	[pp. 4, 26]
{ }	construct and subscript cell array	[p. 53]
3.14	decimal point	[p. 4]
A.B	member of struct or object	[pp. 53, 58]
A.(B)	dynamic member of struct or object	[p. 55]
..	parent folder	[p. 8]
...	continue statement on next line	[p. 4]
,	array rows, parameters, statements	[pp. 4, 13]
;	array columns, terminate statement	[pp. 4, 13]
%	comment	[p. 42]
%%	code cell	[p. 43]
%{	start block comment	[p. 43]
%}	end block comment	[p. 43]
!	system command	[p. 103]
's'	character string	[p. 18]
@	object class folder, function handle	[pp. 61, 67]
~	unused argument	[p. 42]
<	inheritance	[p. 62]

F.4 Strings

F.4.1 String Creation

blanks creates a string of blanks.

cellstr creates a cell array of strings from a character array. See p. 185.

char converts an array of integers to a character array. [p. 101]

sprintf creates a formatted string. See p. 185.

strcat concatenations strings horizontally. `strcat(s1,s2)` is like `[s1,s2]`, except that `strcat` removes trailing spaces.

F.4.2 String Identification

isa tests whether a variable is from a given class (`isa(s,'char')`, for example). See p. 188.

iscellstr tests if a variable is a cell array of strings.

ischar tests if a variable is a string (a char array). Same as `isa(x,'char')`. [p. 70]

isscalar tests if a variable is a scalar. Try `isscalar('a')` and `isscalar('ab')`.

isspace tests if a variable is whitespace (space, tab, return, ...).

isvector tests if a variable is a vector. See p. 170.

F.4.3 String Manipulation

deblank removes trailing blanks from the end of a string. `deblank(c)` for a cell array of strings operates on all strings in c.

lower converts a string to lower case. `lower(c)` for a cell array of strings c operations on all the strings in c.

strrep finds and replaces substrings. See p. 185.

strtrim removes leading and trailing spaces.

upper converts a string to upper case. upper(c) for a cell array of strings c operations on all the strings in c.

F.4.4 String Parsing

findstr Finds strings within another string. Use strfind instead.

regexp searches a string for a regular expression. See p. 185.

regexprep finds and replaces a substring using a regular expression.

sscanf reads formatted data from a string, with a format string similar to the one used by fprintf.

strfind determines if one string is a substring of another string. See p. 185.

strtok extracts tokens from a string. Try [t,r]=strtok('this is a string'), which gives the token t='this' and the remainder r=' is a string'.

F.4.5 String Evaluation

eval executes MATLAB expressions or statements from a string. Avoid whenever possible. See p. 198.

evalin evaluates MATLAB expressions or statements from a string in the workspace of the caller, or in the base workspace. Avoid whenever possible. See p. 198.

F.4.6 String Comparison

strcmp compares strings. strcmp(a,b) is 1 if they are identical, 0 otherwise. To compare multiple strings, one or both arguments can be a cell array of strings. [p. 19]

strcmpi is a case-insensitive version of strcmp. [p. 19] `strmatch` determines if one string is a prefix of another. `strmatch(s,c)` finds the strings in the character array or cell array of strings `c` that have the string `s` as a prefix. With a third argument `'exact'`, only exact matches are sought.

`strncmp` compares the first `n` characters of strings. Otherwise identical to `strcmp`.

`strncmpi` is a case-insensitive version of `strncmp`.

F.5   Bit-Wise Operations

Bit-wise operations can be performed only on unsigned integers, or arrays of unsigned integers.

`bitand` computes a bitwise *and*.

`bitcmp` computes a bitwise *not*.

`bitget` extracts a bit. `bitget(A,1)` is `1` if `A` is odd.

`bitor` computes a bitwise *or*.

`bitset` sets a bit.

`bitshift` shifts bits left or right.

`bitxor` computes a bitwise *xor*.

F.6   Logical Operations

`all` is the $\forall$ logical quantifier. `all(x)` is true if all entries in the vector `x` are nonzero. `all(A)` for a matrix computes a result for each column. [p. 51]

`and` is the logical *and*. `and(A,B)` is `(A&B)`. [p. 16]

`any` is the $\exists$ logical quantifier. `any(x)` is true if any entries in the vector `x` are nonzero. `any(A)` for a matrix computes a result for each column. [p. 51]

`false` returns logical `0`, for false, or an array of logical `0`'s. Its syntax is essentially the same as `zeros`. [p. 17]

find returns the indices and values of nonzeros in a matrix. See p. 180.

logical converts an array to logical. logical(A) is (A~=0). [p. 31]

not is the logical *not*. not(A) is (~A). [p. 16]

or is the logical *or*. or(A,B) is (A|B). [p. 16]

true returns logical 1, for true, or an array of logical 1's. Its syntax is essentially the same as ones. [p. 17]

xor is the logical *xor*. xor(A,B) is (A|B)&(~(A&B)), which is true if either A or B are true, but not both. [p. 16]

F.7 Relational Operations

eq : equal. eq(A,B) is (A==B). [p. 33]

ge : greater than or equal. ge(A,B) is (A>=B). [p. 33]

gt : greater than. gt(A,B) is (A>B). [p. 33]

le : less than or equal to. le(A,B) is (A<=B). [p. 33]

lt : less than. lt(A,B) is (A<B). [p. 33]

ne : not equal. ne(A,B) is (A~=B). [p. 33]

F.8 Set Operations

Sets are represented as vectors of numbers, characters, or cell arrays of strings.

intersect computes the intersection of two sets. [p. 55]

ismember tests if an element is a member of a set. [p. 55]

setdiff computes the set difference of two sets. [p. 55]

union computes the union of two sets. [p. 55]

unique finds the unique elements of a set. [p. 55]

F.9 Date and Time Operations

`clock` returns the current time as a date vector ([year month day hour minute seconds]). The seconds term has a fractional part; all others are integers.

`cputime` returns the total CPU time (in seconds) used by MATLAB since the application started. Not recommended for performance evaluation, since it can wrap around when the internal representation overflows. [p. 73]

`date` returns today's date as a string (excluding the hours, minutes, and seconds).

`datenum` converts a date string to a serial date number. datenum is the inverse of datestr. Try datenum(date) and datenum('1-Jan-0000').

`datestr` converts the result from clock or now into a string. Try datestr(now).

`etime` returns the elapsed time in seconds between two date vectors (as returned by clock).

`now` returns the current time as a serial date number (the number of days since December 31, 2BC, in the Gregorian calendar).

`tic` starts a stopwatch timer. Find the elapsed time of a statement with tic; *statement;* toc. For multiple timers, use s=tic with t=toc(s), where t is the time since s was started. [p. 73]

`toc` reads a stopwatch timer. t=toc gives the time since the last tic. [p. 73]

F.10 Programming in MATLAB

F.10.1 Functions and Scripts

`echo` controls the display of statements as they are executed.

`end` defines the last line of a block of code. See p. 200.

`function` is the first line of any M-file function. `function [a,b]=f(x,y)` declares a function with two inputs x and y and two outputs a and b. [p. 39]

`input` prompts the user for keyboard input. [p. 72]

`inputname` returns the name of an input parameter to a function, as a string.

`mfilename` returns the filename of the currently-running function.

`nargchk` checks if the number of input arguments is valid.

`nargin` is the number of input arguments passed to a function. [p. 41]

`nargout` is the number of output arguments expected from a function. `nargout` is zero if the result is `ans`. [p. 41]

`nargoutchk` checks if the number of output arguments is valid.

`varargin` specifies a variable input argument list. In the `function` statement of an M-file, use `varargin` as the last input argument to collect an arbitrary number of arguments. These are placed in a cell array of the same name. [p. 41]

`varargout` specifies a variable output argument list. In the `function` statement of an M-file, use `varargout` as the last output argument. Then inside the function, assign outputs to `varargout{1}`, `varargout{2}`, and so on. [p. 42]

F.10.2 Evaluation

`ans` returns the result of the most recently evaluated MATLAB expression that was not an assignment. [p. 4]

`builtin` executes a built-in function, ignoring overloading and shadowing. Try `i=3; builtin('i')` and then `which i -all`.

`cellfun` applies a function to each entry of a cell array.

`echo` controls the display of statements as they are executed.

eval executes MATLAB expressions or statements from a string. Try eval('x=4'). Avoid using eval whenever possible. eval can have nasty side effects that are difficult to avoid.

evalin evaluates MATLAB expressions or statements from a string in the workspace of the caller, or in the base workspace (what you see in the Workspace window). For example, evalin('base','x') returns the value of x in the base workspace. Like eval, avoid whenever possible.

feval evaluates a function via a function handle. See p. 187.

pause pauses the Command Window output. Hit any key on the keyboard to continue. pause(3) pauses for 3 seconds. [p. 72]

run runs a script that need not be on the current path.

F.10.3 Timer

timer creates a timer object, which calls a function f(t,e,s) at given intervals (t is the timer object, e is the event, and s is optional). Try t=timer('TimerFcn',@(t,e)why,'Period', 1,'ExecutionMode','fixedRate'). Then start(t) creates a timer that calls why once each second.

delete deletes a timer, with delete(t).

start starts a timer. Try start(t).

stop stops a timer. Try stop(t).

F.10.4 Variables and Functions in Memory

ans returns the result of the most recently evaluated MATLAB expression that was not an assignment. [p. 4]

assignin assigns a value to a variable in the base workspace or caller workspace. Avoid if possible, since it can lead to hard-to-debug code. See also evalin (p. 198).

global declares variables as global. Global variables can be accessed from any function without the need to pass them as arguments. Try not to use global, since it can lead to hard-to-debug code. [p. 39]

`persistent` defines a variable local to a function, which is retained between subsequent calls to that function.

F.10.5 Control Flow

`for` loops across a block of statements terminated with end. for i=a, *statements*, end iterates across the statements, setting the loop index i to one column of a at a time. Try for i=1:5,i,end and for i=eye(3),i,end. [p. 44]

`if` defines a conditional statement. Use with else, elseif, and end. if(A) for a matrix A is true only if *all* entries are nonzero. Try if(x<0), s=-1, elseif(x>0), s=1, else, s=0, end, which is s=sign(x) for the scalar x. [p. 47]

`elseif` is a part of a conditional statement. The elseif(*expression*) is tested only when the if test is false. [p. 48]

`else` is a part of a conditional statement. The else part is executed when the if test and any elseif tests are all false. [p. 48]

`return` is the normal return from a function. A function also returns normally after executing the last statement in the function. Unlike many programming languages, the return statement in MATLAB does not accept any input arguments. [p. 40]

`switch` selects among cases, depending on a single expression. Each case is defined by a case statement. [p. 48]

`case` defines a block of statements for one case of a switch statement. The statements after case(e) are triggered if the switch expression s is equal to e. The statements after case{e1,e2} are triggered if s is equal to e1 or e2. [p. 48]

`otherwise` defines a part of a switch statement. if no case in a switch statement matches the expression, the otherwise block of statements is executed. [p. 48]

`while` executes a block of statements. while(e) tests the expression e each iteration. If the expression e is false, the loop terminates. Try x=2^10, while(x>1), x=x/2, end. [p. 46]

`break` terminates a for or while loop. Statements after the break are not executed. [p. 47]

continue skips the remaining statements in a for or while loop, and goes to the next iteration of the loop. [p. 47]

error abnormally terminates a script or function. error(msg) prints the msg string and terminates the current function. An error may instead be caught via a try/catch statement. [p. 71]

try executes a block of code and recovers from errors. If an error occurs, the try block of code is terminated and the catch block (if any) is executed. [p. 49]

catch defines the second part of a try/catch statement. If an error occurs, the try part is terminated and the statements after the catch part are executed. An optional argument (catch ME) gives information about the error. [p. 49]

end defines the last line of a for, while, if, switch, or try/catch block of code. In a classdef file, end defines the last line of the classdef, properties, methods, and events blocks of code. Also defines the last line of a function, but this is required only if an M-file contains more than one function, or if the function is in a classdef file. [p. 44]

goto does not exist in MATLAB because it leads to unmanageable code. Use break to exit a loop. Use continue to skip a loop iteration. Place code in a function and use return. Use try/catch to handle errors and special conditions. [p. 232]

F.10.6 Error Handling

error abnormally terminates a script or function (see above).

rethrow reissues a previously-caught error. [p. 49]

try executes code and recovers from errors (see above).

catch is the second part of a try/catch statement (see above).

warning issues a warning. warning('oops') displays the warning but does not terminate the function. [p. 72]

F.10.7 MEX Programming

mex compiles a C/C++ or Fortran mexFunction. [p. 87]

G Object-Oriented Programming

G.1 Classes and Objects

class determines the class of a variable. See p. 184.

classdef is a statement that defines a class, with properties and methods. The class definition is terminated by an end statement. [p. 57]

exist determines if a class or object exists. See p. 166.

methods is both a function and a keyword. The function methods(A) lists the public methods of a class or object A. As a keyword, the methods statement defines the set of methods for a class, terminated by an end statement. [p. 59]

properties is both a function and a keyword. The function properties(A) lists the public properties of a class or object A. As a keyword, the properties statement defines the set of properties for a class, terminated by an end statement. [p. 58]

subsref is the method for subscripted references to objects.

G.2 Handle Classes

delete deletes a handle object.

findobj finds a graphics object with specific properties.

get returns the properties of an object. See p. 205.

handle is the superclass for all handle classes, including Handle Graphics. [p. 66]

set lists or sets the properties of an object. set(H) lists the properties that can be modified via set(H,'Property',value). See p. 205.

H Graphics

H.1 Basic Plots and Graphs

[**box**] displays or hides the boundary of a plot.

[**hold**] controls how new data is plotted. With hold on, a plot command overlays the new plot onto the old one. With hold off (the default), a new plot erases the old. [p. 111]

[**line**] creates a line object (a low-level version of plot).

[**loglog**] draws a logarithmic 2-D line plot, with logarithmic x-y axes. Otherwise identical to plot.

[**plot**] draws a 2-D line plot. plot(Y) plots the columns of a real matrix Y versus the row index. With multiple arguments, plot(x,y) plots the vector x versus y. plot(x1,y1,x2,y2,...) plots multiple lines on a single figure. Adding a string after each pair sets the line color and type. [p. 106]

[**plot3**] draws a 3-D line plot. [p. 113]

[**semilogx**] draws a semi-logarithmic 2-D line plot, with a logarithmic x-axis and a linear y-axis. Otherwise identical to plot.

[**semilogy**] draws a semi-logarithmic 2-D line plot, with a linear x-axis and a logarithmic y-axis. Otherwise identical to plot.

[**subplot**] creates a tiled array of plots in a single figure. subplot(2,3,1) creates a 2-by-3 tiling of 6 plots, and sets the first one (in the (1,1) position) as the current plot. [p. 112]

H.2 Plotting Tools

[**rotate3d**] rotates the 3-D view using the mouse. [p. 117]

[**zoom**] zooms into or out of a plot. [p. 107]

H.3 Annotating Plots

legend adds a legend to a plot. If three plot commands have been drawn on a figure, use legend('s1','s2','s3'), where each string describes each plot. Add ,'Location', 'SouthEast' as the last two arguments to place the legend in the bottom right corner (for example). [p. 111]

rectangle draws a rectangle.

title sets the title of a plot. [p. 109]

xlabel sets the label for the *x*-axis. [p. 109]

ylabel sets the label for the *y*-axis. [p. 109]

zlabel sets the label for the *z*-axis, for a 3-D plot. [p. 113]

H.4 Specialized Plotting

area displays an area plot, which is just like a 2-D line plot except that the area below the curve is filled in with a color.

bar draws a bar graph. Try bar(sin(0:.1:pi)).

contour draws a contour plot.

stem plots a discrete sequence of data.

hist draws a histogram plot. hist(y) uses 10 bins; hist(y,20) uses 20. hist(y,x) uses bins centered at the points in x.

histc counts the elements in each bin of a histogram (without actually drawing the plot). The bins are defined differently than hist, however. histc(y,x) defines the bins by their edges, where y(i) is in the kth bin if x(k)<=y(i)<=x(k+1).

fill draws colored polygons.

slice draws a volumetric slice plot. [p. 116]

getframe captures a movie frame.

H.5 Bit-Mapped Images

image displays an image. [p. 119]

imagesc scales and displays an image, so that the full color range is used.

imread reads an image from a graphics file. See p. 169.

imwrite writes an image to a graphics file.

H.6 Printing

orient defines hard-copy landscape/portrait orientation.

print prints a figure.

saveas saves a figure to a file.

H.7 Handle Graphics

H.7.1 Graphics Object Identification

delete deletes a graphics object.

findall finds all graphics objects.

findobj finds a graphics object with specific properties.

gca returns the current axis. [p. 118]

gcbf returns the figure handle of a callback function. [p. 118]

gcbo returns the object handle of a callback function.

gco returns the current graphics object.

ishandle tests if a variable is a Handle Graphics handle.

get gets the properties of a Handle Graphics object. Use get(H) to print all properties, and s=get(H) to return the result as a struct. With the example below (for set), try get(gca,'XTick'), which returns a vector of the tick marks on the x-axis. [p. 118]

set sets the properties of a Handle Graphics object. Use set(H) for a list of properties to set. Try creating a figure (plot(rand(5)), for example). Then use set(gcf), set(gca), and set(gco) to see the properties that can be modified by set. Try set(gca,'XTick',1:5) to change the tick marks on the x-axis, and set(gca,'XGrid','on') to turn on the grid lines on the x-axis. [p. 118]

H.7.2 Object Creation

axes creates a graphics object for the axes of a plot.

figure creates a new figure, or sets the current figure to be a previous-created one. The current figure is where all commands such as plot, title, and xlabel, place their results. figure(3) makes Figure 3 the current figure, creating it if necessary. [p. 107]

image displays an image. [p. 119]

light creates a light at a specified location. [p. 117]

line creates a line object (a low-level version of plot).

patch creates one or more filled polygons.

rectangle draws a rectangle.

surface is a low-level function for creating a surface object.

text adds text to a plot. text(x,y,s) adds the string s to the plot at position (x,y). [p. 109]

uicontextmenu creates a context menu, for right-clicking an object.

H.7.3 Figure Windows

clf clears a figure. clf clears the current figure. clf(2) clears Figure 2. [p. 111]

close closes a figure. close closes the current figure. close(2) closes Figure 2. [p. 111]

closereq is called by default when a figure is closed.

drawnow executes all pending plotting operations. [p. 108]

gcf returns the current figure. [p. 107]

saveas saves a figure to a file.

H.7.4 Axes Operations

axis controls axis scaling for plots. axis auto is the default. axis([xmin xmax ymin ymax]) specifies the limits of the x and y axes. For a 3-D plot, use a vector of length 6. axis tight fits the axes to the data. [p. 110]

box displays or hides the boundary of a plot.

cla clears the current axes.

gca returns the current axis. [p. 118]

grid controls the grid lines of a plot. grid on turns on the grid; grid off turns it off. [p. 109]

ishold tests if hold is on or off. See p. 202.

I 3-D Visualization

I.1 Surface and Mesh Plots

I.1.1 Surface and Mesh Creation

`hidden` removes or reveals hidden lines in a mesh plot.

`mesh` draws a 3-D mesh plot. Try mesh(peaks). [p. 114]

`peaks` is an example function for surface plots. peaks returns a matrix of function evaluations $f(x,y)$ over a uniformly distributed set of x-y points, for a certain function f. [p. 117]

`surf` creates a 3-D surface plot. See the cover of this book for an example. Try surf(mesh). [p. 114]

`surface` is a low-level function for creating a surface object.

I.1.2 Domain Generation

`meshgrid` creates X, Y, and Z arrays for 3-D plots. [p. 114]

I.1.3 Color Operations

`colorbar` adds a color legend to a plot. [p. 117]

`colormap` sets or returns the colormap for an image. [p. 117]

`shading` sets color shading properties for 3-D plot. [p. 116]

I.2 View Control

I.2.1 Camera Viewpoint

`view` specifies the viewpoint for a 3-D plot. [p. 117]

I.2.2 Aspect Ratio and Axis Limits

`xlim` sets or queries the *x*-axis limits.

`ylim` sets or queries the *y*-axis limits.

`zlim` sets or queries the *z*-axis limits.

I.2.3 Object Manipulation

`reset` resets the properties of a graphics object to their defaults.

`rotate3d` rotates the 3-D view using the mouse. [p. 117]

`zoom` zooms into or out of a plot. [p. 107]

I.3 Lighting

`light` creates a light at a specified location. [p. 117]

`lighting` specifies the lighting algorithm (flat, gouraud, phong, or none). [p. 117]

I.4 Volume Visualization

`slice` draws a volumetric slice plot, which plots slices of data from a 3-D volume. [p. 116]

J GUI Development

J.1 Predefined Dialog Boxes

`errordlg` displays an error dialog. Try errordlg('ack!').

`inputdlg` displays an input dialog. Try inputdlg('x:').

`msgbox` displays a message box. Try msgbox('done').

`questdlg` displays a question box. Try questdlg('go?').

`uigetdir` displays a dialog box for selecting a directory.

`uigetfile` displays a dialog box for accessing files.

`uiputfile` displays a dialog box for saving files.

`waitbar` displays a wait bar. Try h=waitbar(0,'working'); and then for x=0:.1:1, waitbar(x,h), pause(1), end.

`warndlg` displays a warning dialog.

J.2 User Interface Deployment

`guidata` stores or retrieves GUI data.

`guihandles` returns all handles in a figure.

`movegui` moves a GUI figure to a specified screen location.

`openfig` opens a figure from a .fig file.

J.3 User Interface Development

`getappdata` gets the value of application-defined GUI data.

`getpref` gets the current setting of a preference.

`ginput` gets graphical input from the mouse or cursor.

`guidata` stores or retrieves GUI data.

`isappdata` tests if application-defined data exists.

`rmappdata` removes application-defined data from an object.

`setappdata` sets the value of application-defined GUI data.

`waitfor` pauses until a specified condition occurs.

`waitforbuttonpress` pauses for the user input.

J.4 User Interface Objects

`menu` creates a GUI menu.

`uicontextmenu` creates a (right-click) context menu.

`uicontrol` creates a user interface object. [p. 118]

`uimenu` creates a user-interface menu.

`uipanel` creates a panel for grouping components.

J.5 Objects from Callbacks

`findall` finds all graphics objects.

`findobj` finds a graphics object with specific properties.

`gcbf` returns the figure handle of a callback function. [p. 118]

`gcbo` returns the object handle of a callback function.

J.6 Program Execution

`uiresume` resumes execution after uiwait. Typically used in a callback function. [p. 118]

`uiwait` pauses execution until uiresume is called. [p. 118]

K External Interfaces

K.1 Shared Libraries

`calllib` calls an external function in a shared library.

K.2 Java

`import` imports a Java package for use in MATLAB. [p. 100]

K.3 Component Object Model and ActiveX

`actxserver` creates a COM server on Microsoft Windows.

`delete` deletes a COM server.

`invoke` invokes a method on a COM server.

L Symbolic Math Toolbox

L.1 Calculus

`diff` performs symbolic differentiation. Try
`syms x; diff(x^2).` [p. 127]

`int` performs symbolic indefinite or definite integration. Try
`syms x; int(x^2);` and `int(x,0,1).` [p. 129]

`limit` computes the limit of a symbolic expression. [p. 131]

L.2 Linear Algebra

`det` computes the determinant of a symbolic matrix. [p. 144]

`diag` extracts the diagonal of a symbolic matrix, or creates a symbolic diagonal matrix. See p. 171.

`eig` computes the eigenvalues and eigenvectors of a symbolic matrix, or finds the generalized eigenvalues/eigenvectors. [p. 143]

`inv` computes the inverse of a symbolic matrix. Does not suffer the inaccuracy of `inv(A)` for a numerical matrix A. [p. 143]

`null` computes a basis for the null space of a symbolic matrix.

`poly` constructs a polynomial from a set of specified roots, or constructs the characteristic polynomial of a matrix. [p. 143]

`rank` computes the rank of a symbolic matrix.

`svd` computes the singular value decomposition of a symbolic matrix. [p. 143]

`tril` extracts the lower triangular part of a symbolic matrix. See p. 173.

`triu` extracts the upper triangular part of a symbolic matrix. See p. 173.

L.3 Simplification

factor factors a symbolic expression into a product of subexpressions, or factors a numeric or symbolic nonnegative integer into its prime factors. [p. 135]

subs substitutes one symbolic expression into another. subs(s,2) replaces the default variable (typically x) with the value 2. subs(s,y,z+1) replaces y with z+1 in the expression s. All occurrences are replaced. [p. 133]

L.4 Special Functions

zeta computes the Riemann zeta function, $\zeta(x)$, or its derivatives.

L.5 Conversions

double converts a symbolic expression to a numeric one (in IEEE double precision). [p. 131]

L.6 Basic Operations

ceil rounds a symbolic expression towards $+\infty$. See p. 178.

conj computes the complex conjugate of each entry in a symbolic array.

eq tests if two symbolic expressions are equal. eq(A,B) is (A==B). See p. 195.

fix rounds a symbolic expression towards zero. See p. 178.

floor rounds a symbolic expression towards $-\infty$.

imag returns the imaginary part of a symbolic expression.

log10 computes the base-10 logarithm of a symbolic expression.

log2 computes the base-2 logarithm of a symbolic expression.

mod computes the remainder after symbolic division. See p. 178.

real returns the real part of a symbolic expression. [p. 153]

round rounds a symbolic expression towards the nearest integer.

size returns the size of a symbolic array. See p. 171.

sort sorts entries in a symbolic array, as if they were text entries.
sort([z 1 y x]) is [1, x, y, z].

sym creates a symbolic number or variable. sym(A) where A is a numeric matrix constructs a symbolic representation of the matrix. A second argument controls how the numbers are converted ('f' for floating-point, 'r' for rational, and 'd' for decimal). sym('x') creates a symbolic variable. A second argument controls what kind of symbolic variable is created ('real' if x is real, 'positive' if x is real and positive, and 'clear' if there are no restrictions on x).

syms creates symbolic numbers and variables. syms x y is short for x=sym('x'); y=sym('x'). [p. 126]

Index

C Fever, by Tim Davis

~

I must go code in both C and M, not only C and VI,
And all I ask is a Linux box and a mouse to steer her by,
And the `while`'s break and the `if-then`
      and the `valgrind`'s shaking,
And a dash `O` so the C's fast and a `switch case break`-ing.

I must go code in both C and M,
      for the call of recursive code
Is a wild call and a clear call that never will be slowed;
And all I ask is a latte (tall, with the white foam frothing),
And no flung err and no blown stack,
      while the C goes flying.

I must go code in both C and M, to the lonely code frontier,
Where no `goto`'s in the `while`'s way,
      where the logic is sharp and clear;
And all I ask is a `#define` from a fellow C-programmer,
And quiet `lint` and a clean `doc` in the LaTeX grammar.

~

Based on *Sea Fever*, by John Masefield (1902). *"I must go down to the seas again, to the lonely sea and the sky, and all I ask is a tall ship and a star to steer her by. ..."*

`#define` is pronounced as *pound define*. LaTeX sounds like *la tech*. `valgrind` rhymes with *grinned*. The `-O` option enables code optimization for the `mex` command and for typical C compilers.

232